Einstein, Tagore and the Nature of Reality

The nature of reality has been a long-debated issue among scientists and philosophers. In 1930, Rabindranath Tagore and Albert Einstein had a long conversation on the nature of reality. This conversation has been widely quoted and discussed by scientists, philosophers and scholars from the literary world. The important question that Tagore and Einstein discussed was whether the world is a unity dependent on humanity, or if the world is a reality independent of the human factor. Einstein took the stand adopted by Western philosophers and mathematicians, namely that reality is something independent of the mind and the human factor. Tagore, on the other hand, adopted the opposite view. Nevertheless, both Einstein and Tagore claimed to be realists despite the fundamental differences between their conceptions of reality. Where does the difference lie? Can it be harmonised at some deeper level? Can Wittgenstein, for example, be a bridge between the two views? This collection of essays explores these two fundamentally different conceptions of the nature of reality from the perspectives of theories of space-time, quantum theory, general philosophy of science, cognitive science and mathematics.

Partha Ghose is an Honorary Scientist at the National Academy of Sciences, India.

Routledge Studies in the Philosophy of Mathematics and Physics

Edited by Elaine Landry, University of California, Davis, USA and Dean Rickles, University of Sydney, Australia

1 **The Logical Foundation of Scientific Theories**
Languages, Structures, and Models
Décio Krause and Jonas R.B. Arenhart

2 **Einstein, Tagore and the Nature of Reality**
Edited by Partha Ghose

Einstein, Tagore and the Nature of Reality

Edited by Partha Ghose

Routledge
Taylor & Francis Group
LONDON AND NEW YORK

First published 2017 by Routledge

2 Park Square, Milton Park, Abingdon, Oxon, OX14 4RN
605 Third Avenue, New York, NY 10017

Routledge is an imprint of the Taylor & Francis Group, an informa business

First issued in paperback 2020

British Library Cataloguing in Publication Data
A catalogue record for this book is available from the British Library

Library of Congress Cataloging-in-Publication Data
Names: Ghose, P. (Partha), editor.
Title: Einstein, Tagore, and the nature of reality / edited by Partha Ghose.
Description: 1 [edition]. | New York : Routledge, 2016. | Series: Routledge studies in the philosophy of mathematics and physics ; 2 | Includes bibliographical references and index.
Identifiers: LCCN 2016008565 | ISBN 9781138685246 (hardback : alk. paper)
Subjects: LCSH: Reality. | Einstein, Albert, 1879–1955. | Tagore, Rabindranath, 1861–1941. | Science—Philosophy.
Classification: LCC BD331 .E425 2016 | DDC 111—dc23
LC record available at http://lccn.loc.gov/2016008565

ISBN: 978-1-138-68524-6 (hbk)
ISBN: 978-0-367-73701-6 (pbk)

Typeset in Sabon
by Apex CoVantage, LLC

Contents

Figures

Preface

The nature of reality has been a long-debated issue among scientists and philosophers. Albert Einstein (1879–1955) invited Rabindranath Tagore (1861–1941) to his house in Kaputh, Germany, on 14th July, 1930, and had a long conversation with him on the nature of reality. This conversation has been widely quoted and discussed by scientists, philosophers and scholars from the literary world.

The important question that Tagore and Einstein discussed was whether the world is a unity dependent on humanity, or if the world is a reality independent from the human factor. Einstein took the stand adopted by Western philosophers and mathematicians such as Aristotle, Plato, Frege, Russell etc., namely that *reality* is something independent of the mind and the human factor. On the other hand, Tagore adopted the opposite view. Nevertheless, both Einstein and Tagore claimed to be realists, but their conceptions of reality were obviously fundamentally different. Where does the difference lie? Can it be harmonized at some deeper level? Can Wittgenstein, for example, be a bridge between the two views?

There are no universally agreed-upon definitions of truth and reality in Western philosophy. Realisms of various shades of meaning have developed over the years, such as naïve realism, representative realism, direct realism, transcendental realism, Platonic realism, moderate realism, new realism, organic realism, constructive realism, entity realism etc.

In the philosophy of science, the focus has been more on to what extent the world described by science is the real world (critical or scientific realism). Classical science was clearly realistic in nature in the sense of being observer independent and deterministic. Quantum mechanics changed this perspective. Whether or not the world described by quantum mechanics is real and deterministic has been hotly debated since its inception. The latest predominant view is that it is not. Since all experiments can give us is information about the world, there is a growing school of thought that looks at the world as a giant information processor and prefers to look at information itself as reality. What implications does this emerging view have on the perennial debate?

In classical Indian philosophy, too, this debate was carried out with amazing analytical skills, leading to developments in the Samkhya-Yoga worldview (a dualist view), Kashmir Shaivism (a monistic view), Buddhist philosophy (ranging from straightforward realism through the Middle Path to mind-only idealism), Jaina realism and *anekantavada*, *Nyaya* realism and the *Vedanta* schools varying from dualism through qualified non-dualism to strict non-dualism (*advaitavada*). Each school has its own subtlely nuanced concept of reality. Tagore developed his own view of reality based on Indian philosophy broadened by his knowledge of Western philosophy and science and his own spiritual experience. How does it stand in relation to current developments in science and philosophy?

If we are concerned with the purpose of life, what we should be doing and the *meaning* of life and death, then the issue of reality (what *is*) and ethics (what *ought* to be) is of crucial importance to us. If things are different from how they are presented to us, then the issue of reality becomes crucial to our survival and well-being. Hence, the nature of reality has been an important concern in literature, art, philosophy and the social sciences. How do these concerns relate to the Einstein-Tagore debate?

It has long been realized that the mind plays a fundamental role in recognising and characterising reality. Hence, knowledge of the character of the mind is crucial to understanding the nature of reality. Recent progress in understanding the evolution of language and advances in neuroscience and cognitive science have led to many discoveries with important bearings on the debate that need to be raised and discussed.

An impression has been created that the Einstein-Tagore debate was a failure, perhaps because of the following account of the event written by Dmitri Marianoff, Einstein's step son-in-law that appeared in *The New York Times* issue of 10 August, 1930: 'Einstein listened with studious attention, then gave his characteristic view. Neither sought to press his opinion. They simply exchanged ideas. But it seemed to an observer as though two planets were engaged in a chat'. There is a parallel in the history of quantum mechanics, namely, the famous Bohr-Einstein debate. Unlike one afternoon's conversation between a scientist and a poet in which language was certainly a barrier (Einstein was not fluent in English and Tagore did not understand German, so there was an interpreter), the Bohr-Einstein debate was between two outstanding scientists of the time over a subject that both helped to create, and it stretched over nearly twenty-eight years (from 1927 to 1955). The subject matter was essentially the same, namely, the role of an observer in describing nature. Bohr pushed his theme of the essential role of measurements (and hence of observers) in describing nature, and Einstein stuck to his faith in an observer-independent reality. Their metaphysical stands were radically different. Physics has benefited from the great debate, which still goes on, and it is hoped that the Einstein-Tagore debate will open up other aspects of the nature of reality and contribute significantly to progress in the philosophy of science as a whole.

To commemorate the 150th birth anniversary of Tagore, a three-day-long international seminar on 'The Nature of Reality: The Perennial Debate' was organised from 1–3 March, 2012, at the Indian Institute of Advanced Study, Shimla. The intention was to have wide-ranging, in-depth discussions on the entire gamut of man's engagement with reality through science, philosophy, language, cognitive science and the social sciences while keeping the Tagore-Einstein conversation as a reference point. This volume is a collection of essays written on the basis of the talks delivered at the Shimla conference as well as some additional ones specially written for this volume. The chapters explore the two fundamentally different conceptions of the nature of reality reflected in the 1930 Einstein-Tagore conversation from the perspectives of (i) theories of space-time, (ii) quantum theory, (iii) general philosophy of science, (iv) cognitive science and (v) mathematics.

Introduction

In the Introduction, Kathleen O'Connell describes the background of the meetings between Einstein and Tagore, the two iconic figures of the 20th century, one a scientist from the west and the other a poet from the east, both at a personal level and within their socio-historical times. It details the dates and locations of their personal meetings in 1926 and 1930. It also explores mutual interests such as science, nature, education and music; shared acquaintances such as the Indian scientist Sir Jagadish Chandra Bose, the discoverer of microwaves and a pioneer in plant electro-physiology, and the French savant Romain Rolland; their perceptions concerning one another and their roles in opposing aggressive nationalism and fascism.

At their first meeting at Einstein's residence in Kaputh on 14th July, 1930, Tagore began the conversation with the words, 'You have been busy hunting down with mathematics the two ancient entities, Time and Space, while I have been lecturing in this country on the eternal world of Man, the universe of Reality.' Hence, it is befitting that after the Introduction, the volume begins with a chapter on the nature of space and time.

Space-time and reality

In Chapter 1, Brown and Lehmkuhl explore Einstein's search for the nature of space and time. In the years between developing the special and general theories of relativity, Einstein came to see flat space-time as a real entity whose causal role violates the so-called action–reaction principle because it remains unaffected by matter, energy and events. He consequently regarded his 1915 theory of gravity—his general theory of relativity—as a triumphant resuscitation of the principle, while Newtonian mechanics as well as his 1905 special theory of relativity supposedly violated it. The authors examine why Einstein came to emphasize this position several years after the development of general relativity. Several key considerations are relevant to the story: the connection Einstein originally saw between Mach's analysis of inertia and both the equivalence principle (the exact equality of inertial and gravitational masses and the consequent local equivalence of gravitation and accelerated reference frames) and the principle of general covariance, the waning of Mach's influence owing to de Sitter's 1917 results and Einstein's detailed correspondence with Moritz Schlick in 1920.

In Chapter 2, Cao explores the nature or the truth of space and time from a perspective based on a constructive version of *structural realism* that appears to be a synthesis of the views espoused by Tagore and Einstein. In their conversation, Einstein's clearly stated view that 'Truth must be conceived as a Truth that is valid independent of humanity' is in sharp contrast with Tagore's equally clearly stated view that 'the Truth of the Universe is [can only be] human Truth.' Like Brown and Lehmkuhl, Cao also devotes considerable attention to Einstein's deliberations on the so-called 'hole argument' in the years leading to his 1915 formulation of the general theory of relativity and their philosophical implications, especially those concerning the nature of space and time. Cao examines these in the context of the dominant views on space and time held by Newton, Leibniz, Kant, Engels, Marx and those suggested by recent developments in the studies of quantum gravity. Cao's examination seems to show that the reality of space-time is the result of stepwise human construction: from the very abstract Kantian form of pre-space-time to the more and more concrete reality of space-time, which seems to be in line with Tagore's wisdom. On the other hand, although these constructive steps can be viewed as steps in the ascendance from the abstract to the concrete in Hegel's dialectical logic, which was used by Karl Marx in his construction of social reality, the constructive ascendancy described in this chapter involves not the logical categories, but physical entities, events and processes, and thus this dialectical constructive approach to understanding the nature of space-time should be regarded as an application of Marx's dialectics rather than Hegel's; that is, as a human construction based on objective reality, which seems to be in line with Einstein's 'religious belief in super-human objectivity.'

Quantum mechanics and reality

No discussion on reality can be complete without delving into the implications of quantum theory. Accordingly, in Chapter 3, Charles H. Bennett explains how quantum physics has helped explain the nature of information and the origin of randomness. In the early part of the 20th century, quantum mechanics suddenly opened up a strange new world of counterintuitive simplicity that physicists and philosophers are still digesting. Although originally devised to explain the behaviour of tiny objects like atoms and photons, quantum theory is fundamental to the understanding of nature on all scales. In addition to having withstood a century of experimental tests, it provides a picture of reality that is ultimately more satisfying and coherent than either the strict determinism of Laplace or the blatant probabilism that so repelled Einstein about quantum mechanics as he understood it. Bennett sketches, in simple, technical terms, the subtler understanding of quantum principles developed since Einstein's death, emphasising the central role of *entanglement*, a peculiarly intense and unsharable kind of correlation made possible by quantum laws that Einstein himself was the first to recognize, and *decoherence*, a loss of correlation within a system that occurs when the system becomes entangled with its environment. From the vantage point of entanglement and decoherence theory, Bennett takes us through the basic laws of the quantum world to give us a glimpse of the origin of quantum randomness in entanglement, the difference between classical and quantum information processing, quantum cryptography, the ambiguity of the past and the future, quantum Darwinism, the threat of the invasion of privacy and the ontological status of escaped and lost information, to point out why the question, 'Why did this happen instead of that?' is sometimes no more answerable, nor requiring of an answer, than the question, 'Why am I me instead of someone else?', that we all learned to stop asking as children. In the end, he points out that according to many cosmologists, some as-yet-unexplained accidental features of the universe (such as the small-but-nonzero cosmological constants or the anisotropy of the cosmic microwave background) probably had their origin in a Schopenhauer form of the weak anthropic principle, and that 'it is unreasonable to deny that some sort of anthropic selection may be biasing our view of the universe.' In other words, we live in 'a world just barely compatible with our existence'. That is his take on the Einstein-Tagore 'debate'.

In Chapter 4, Anthony Sudbery examines the two broad, opposing classes of attitudes to reality (realist vs. idealist, material vs. mental) with corresponding attitudes to knowledge (objective vs. subjective, scientific vs. romantic). According to Sudbery, this clash can be seen at its strongest in Blake's rejection of the scientific view of the world, which he found personified in Newton. Blake's picture of Newton shows him focusing on a geometrical figure and missing the rich reality around him, which can only be found in personal, subjective experience, not in the general, objective view of reality that is the aim of science. It is one of the most enigmatic features of quantum mechanics, the most fundamental and accurate of all sciences, that it is hard to reconcile it with this objective view. Sudbery uses the Everettian 'relative state' or 'many worlds' interpretation to illustrate that in order to understand quantum mechanics, one must heed Blake's lesson and accept that there is a subjective reality—a reality relative to a particular observer inside the world who is also subject to the laws of quantum mechanics—which is as valid as the objective reality seen from outside that world. Each of these kinds of worlds or realities is needed to make sense of the other; they are not contradictory, but compatible and complementary.

In a somewhat similar vein but using different arguments, Ravi Gomatam starts Chapter 5 with the observation that, though taken at face value,

Einstein and Tagore disagreed over the nature of truth or reality, both of them believed in a Reality that humans could comprehend. They only differed over how far the two—Reality and our conception of it—can be synthesized. He argues that, in the context of quantum mechanics, Einstein did not distinguish between a subjective recording of a measurement outcome and the measurement itself, which is an objective interaction in the world, though the distinction is crucial. Tagore, in contrast, held that Truth and Reality in the absolute sense can be comprehended only via a relational but individual perspective. But he limited this perspective to a spiritual search for divinity by the individual, in perfecting the relation with the Universal Person, an idea Gomatam traces to the Bengali *Vaishnava* tradition of the Bauls (a heterodox sect of wandering minstrels who sing of the Man within the mind). With a view to synthesising Einstein's and Tagore's positions so as to bear upon the issues of causality and reality in quantum mechanics, Gomatam introduces the tenets of the original Bengali *Vedanta* tradition, named the *Gaudiya Vaishnava Vedanta* (GVV) and presented by the saint-philosopher Chaitanya over five centuries ago. Chaitanya proposed the philosophy of *achintya bhedaabheda* (inconceivable simultaneous oneness and difference), which is neither monistic nor dualistic, although it integrates features of both. Specifically with reference to the Einstein-Tagore dialogue, Gomatam argues that GVV advances the idea of relation as being ontologically primary, with objects epistemologically derived. This is in contrast to the relational viewpoint that has so far prevailed in physics since early mechanics, in which objects are ontologically primary and relations are epistemologically derived. He discusses how this new ontological relational viewpoint of matter could be appropriate to address the nature of quantum reality and causality.

General philosophy of science

The following chapters deal with these questions from a more general philosophical point of view. In Chapter 6, Tushar K. Sarkar starts with the observation that the question, 'What is Reality?' is mainly meta-scientific or philosophical in nature rather than a purely scientific one. Accordingly, he begins by unpacking the different strands of the question: epistemic, ontological, conceptual and methodological. Since the answer to a meta-scientific question depends more on an expert's methodological leaning than on the results of scientific experiments, and a scientist's methodological leaning depends, in part, on the ruling paradigm(s) of the time, the answer to a question is often determined in practice by a mutual feedback loop between a methodology and an accepted paradigm. This often gives rise to conflicting answers to a question. Sarkar provides some illustrative examples. He then gives a summary of the salient features of the standard version of quantum mechanics in order to identify the exact issues that lie at the root of the controversies between Bohr and Einstein, and to give a definite shape to the question of what Reality is. He does this by raising some meta-scientific questions and pointing out that a scientific theory is no more than a bunch of symbols until they are interpreted. The conceptual-philosophical issues raised during the Einstein-Tagore conversation are then put in their contexts by looking at each one's (i) scientific and (ii) socio-cultural backgrounds, which reveal some affinities between Tagore's view and the quantum mechanical view of reality. Keeping in mind Tagore's grounding in the *Upanishads* (the later and philosophical parts of the ancient Indian *Vedas*, which contain Sanskrit verses about the ultimate reality called *Brahman*), which regard truth/reality-consciousness-bliss as a holistic unity, he then discusses, in a somewhat different vein than Gomatam, the plausibility of introducing consciousness as a legitimate explanatory parameter, matter-consciousness relationships, ways of bridging the gap between direct sensory inputs and the knowledge constructs constrained by our built-in knowledge representation mechanisms and gives some reasons for busting the myth of pure objectivity. Finally, he introduces the distinction between a photographic snapshot, a portrait-painting and a creative, imaginative 'vision' of the nature of Reality, and argues why a creatively imaginative reconstruction of Reality that is neither a snapshot nor a portrait is the *only* plausible option for us, scientists and poets alike.

In Chapter 7, C. S. Unnikrishnan examines the interplay between 'observable' and 'unobservable' and 'insensible' entities that inevitably occurs in theory construction. Entities like fields, wave functions and even space and time are all unobservables except as manifestations of material existence and behaviour. There is thus an obvious difference between the reality associated with these unobservable theoretical entities and that of perceptible matter. The success of the physical theory is often taken as evidence for the physical reality of such unobservables. While a rigorous natural philosophy cannot take them for granted, there does not seem to be a way of avoiding such unobservables if one has to construct theories. From this point of view, Unnikrishnan makes the case that physical theories are a 'constructed' reality, consistent observations. While observables and unobservables coexist peacefully in most of classical physics, conflicts and dissonance arise when microscopic physics is described by quantum mechanics. Cosmology also provides an observational framework to bring out some of these conflicts. Unnikrishnan gives three examples. One is about the loss of physical reality due to the need for a consistent quantum mechanical representation. The second example deals with the conflict between the assumed reality of quantum fields, so fundamental and essential to our standard physics worldview, and the observable universe. The third deals with an all-important difference between conventional modern physics constructed in the unreal, nearly empty 'space' and an empirically stronger physics with matter-filled universe as its arena. Not surprisingly, the latter, with its universal cosmic links and physical inseparability, also holds human concepts of harmony and beauty.

In Chapter 8, K. Sridhar views the Tagore-Einstein discussion as providing several vantage points from which to address questions of scientific truth, its epistemology and methodology, empiricism and realism. In the last few years, a vigorous case has been made by some philosophers of science and scientists for revisiting the question of empiricism in science. It has been argued that the classical form of empiricism based on verification or falsification, which is still the basis for all discussions of scientific methodology, is no longer a viable methodology and needs to be replaced by a less demanding and a possibly probabilistic one. This has been strongly criticized by scientists who treat empiricism, of the traditional kind, as the hallmark of the scientific method. Tied intimately to this debate is the question of realism, especially in the context of modern-day atomism. Sridhar looks at atomism and realism within the context of these debates.

Reality and cognitive science

Cognitive science is an emerging multidisciplinary approach that offers a completely different perspective on realism. In Chapter 9, Avi Chaudhuri offers a clear perspective from this point of view. He begins by asking: If a tree falls in a forest and no one is there to hear it, will it still create a sound? This famous question highlights a central debate surrounding two different conceptions about the nature of reality: the world as it exists independent of humanity versus the world as a mental construct that is solely dependent on human perception. Of course, the falling tree would make a sound, according to Einstein, the physicist. However, Tagore, the poet and philosopher, would argue to the contrary because reality according to him is in turn dependent upon human consciousness. Chaudhuri follows Tagore's thesis from the perspective of cognitive neuroscience and argues that even cases of conscious perception rely on various enigmatic facets of the human mind, such as *attention*, *motivation*, *emotion* and *culture*. The biological basis of attention has been a particular area of intense study because of its gating functions in sensory information transfer to the mind. He discusses this phenomenon with illustrations and examples through everyday phenomena, along with some captivating illusions that show a striking failure to apprehend reality. He ends by asking the question: If a human being is in the midst of reality but does not perceive it, does that reality still exist?

Chapter 10 has three main parts. The first part expounds in some detail the different concepts of reality held by Einstein and Tagore, also explored by other contributors. The second part deals with the recently empirically proven occurrence of *entanglement* (so far believed to be a quintessentially quantum phenomenon) in *classical optics*, shifting the boundary between the quantum and classical domains and sharpening the debate on the nature of reality emerging from modern physics. The third part is concerned with an alternative approach to reality based on *cognitive science*, introduced in the previous chapter, which is usually ignored by physicists. The debate between the theses of *embodied cognition* and *artificial intelligence* is outlined. The former holds that the nature of the 'mind' is largely determined by the form of the body, i.e., the particular way in which an organism is embodied (e.g., whether it has feet, fins, eyes, a tail etc.) will influence how it performs goal-directed actions in the world, and the particular sensorimotor experiences connected with these actions serve as the basis for category and concept formation. In other words, cognition is essentially *constructive*, something that Cao, Chaudhuri and Sarkar also emphasize. This is in contrast with the approach of artificial intelligence, which uses the *computer metaphor* of the mind and views cognition as a passive retrieval. Tagore clearly believed in embodied cognition, while Einstein appears to have favoured the computer metaphor of the mind.

Mathematical reality

Mathematics underlies the basis of all quantitative sciences, and its importance can be traced back to the ancient times of Pythagoras ('All is numbers'), Aristotle ('The principles of mathematics are the principles of things'), Plato, who insisted that no one should enter his Academy who has no knowledge of geometry, and the ancient Indian astronomical text *Surya Siddhanta*, whose antiquity cannot be established but certainly goes back to the Hellenistic Greek period when there was cultural contact between Indian and Greek astronomers. Mathematics was used and developed further in India by the mathematician-astronomers Aryabhata (476–550 CE), his contemporary Varahamihir and later mathematicians and commentators Brahmagupta and Bhaskara I, and in Europe by Galileo ('The book of nature is written in the language of mathematics') and Newton. However, with the ever-increasing abstractness and precision of physical theories like relativity and quantum mechanics in the twentieth century, there has been a growing disconnect between the reality portrayed by the mathematically underpinned physical sciences and the reality of direct experience.

In 1931, when Bertrand Russell, Alfred North Whitehead and David Hilbert were using logic to understand the foundations of mathematics, Kurt Friedrich Gödel made an enormous impact on scientific and philosophical thinking by publishing his famous *incompleteness theorems*. He proved that in any computable axiomatic system that is powerful enough to describe the arithmetic of natural numbers, (i) a consistent system cannot be complete and (ii) the consistency of the axioms cannot be proven within the system. This ended the programme started by Gottlob Frege culminating in Russell and Whitehead's *Principia Mathematica* and Hilbert's *formalism*. In recent years, Roger Penrose has used these incompleteness theorems to argue that the human mind has abilities that are non-computable and which no Turing machine (a hypothetical device representing a computer) can possess. In other words, the human mind is non-algorithmic. One does not, of course, know what form a non-computable science would have, but one thing is certain: It would be quite unlike present-day science. This reopens the debate on the nature of mathematical reality and its relationship with the human mind.

In the final chapter, Mihir K. Chakraborty offers a glimpse of the mathematician's view of reality. He touches on Gödel's views on the philosophy of mathematics and gives a brief introduction to Gödel's incompleteness theorems in a historical perspective. He also offers his own understanding of the philosophical issues involved and his realisation in this regard of the significance of the Einstein-Tagore conversations.

Acknowledgement: We acknowledge the National Academy of Sciences, India, for grants that enabled the present work to be undertaken.

Partha Ghose, Editor

Introduction

Kathleen M. O'Connell

The relationship and exchange of ideas that took place between two of the most iconic personalities of the twentieth century, Rabindranath Tagore (1861–1941) and Albert Einstein (1879–1955), has provided an area of continued interest over the years. This volume, which concentrates on the scientific and philosophical implications of their exchanges, attests to that continuing interest. The introduction will examine them at a more personal level, exploring their interests, acquaintances and perceptions concerning one another, as well as their roles in the broader socio-historical context.

On the surface, it would seem as though the two figures hold little in common. There is a well-circulated and mesmerising photograph taken of the two in Germany in 1930, which seems to accentuate their differences. The two Nobel Laureates look out at us with penetrating eyes. Tagore, elegant and poised in a *rishi*-like textured robe, looks directly at us with intense expression that seems to reflect deep consideration of a profound problem. By contrast, Einstein, with hair rather wildly unkempt and in a dark suit with tie slightly askew, is looking slightly upward, giving the appearance of contemplating something in the distance in somewhat abstract fashion. Describing the two, Dmitri Marianoff, who married Margot Einstein (stepdaughter of Albert) in November 1930, offered the frequently quoted comment: 'Tagore, the poet with the head of a thinker, and Einstein, the thinker with the head of a poet'.

Despite obvious differences, a closer examination of the two individuals within their historical and cultural times reveals a number of mutual affinities and concerns. There is, as mentioned, their shared roles as Nobel Laureates. Tagore was awarded the Nobel in Literature in 1913 (the first non-European to do so), *"because of his profoundly sensitive, fresh and beautiful verse, by which, with consummate skill, he has made his poetic thought, expressed in his own English words, a part of the literature of the West"*. Einstein received the Nobel in 1921 in Physics *"for his services to Theoretical Physics, and especially for his discovery of the law of the photoelectric effect"*. Both individuals, apart from their particular fields, were caught up in the profound social and historical changes that were affecting peoples' lives everywhere. They each expressed concern over the rise of aggressive nationalism and became champions of human rights and creative freedom in the pursuit of world peace. Beyond the scientific and political dimensions, they shared an interest in educational trends and the arts, particularly music.

Tagore's political activities had included the organising of political opposition against the 1905 partition of Bengal and his writings, such as the powerful political novel *Home and the World (Ghare-Baire)* published in 1917, explore the political and socio-religious effects of narrow sectarianism. He was deeply affected by the First World War and condemned war and militarism in his book *Nationalism,* also published around the same time. Einstein was vocal in his opposition to German militarism and gave up his citizenship at the age of 17, refusing military service and moving to Switzerland to study at the Zurich Polytechnic. He returned to the University of Berlin later, reacquiring German citizenship and continued his anti-war activities within Germany.

Their shared concerns about national chauvinism brought them into contact with Romain Rolland, another Nobel Laureate (Literature, 1915), who invited each to sign an anti-nationalist document, 'declaration pour l'independence de l'esprit' in 1919, to which they both agreed. Other signers included: Jane Addams (USA), Tolstoy's secretary Pavel Birukov (Russia), Benedetto Croce (Italy), Georges Duhamel (France), Hermann Hesse (Germany), Selma Lagerlof (Sweden) and Bertrand Russell (England), among others. At the time, Tagore had replied to Romain Rolland's request in an open letter dated July 9, 1919:

> When my mind was steeped in the gloom of the thought, that the lesson of the war had been lost, and that people were trying to perpetuate their hatred and anger into the same organised menace for the world which threatened themselves with disaster, your letter came and cheered me with its message of hope. The truths that save us have always been uttered by the few and rejected by the many and have triumphed through their failures. It is enough for me to know that the higher conscience of Europe has been able to assert itself in one of her choicest spirits through the ugly clamours of passionate politics; and I gladly hasten to accept your invitation to join the ranks of those freed souls, who in Europe have conceived the project of a Declaration of Independence of the Spirit.[1]

Jagadish Chandra Bose, scientist and close friend of Tagore's, was another figure linking these individuals in their scientific interests, concerns about education and internationalism. Tagore had helped raise money to send him to England to continue his work in physics and botany. Bose was in close touch with Rolland and a member of the 1921 Committee on Intellectual Cooperation, formed by the League of Nations, which also included Einstein, Henri Bergson, Eve Curie and Gilbert Murray. The committee outlined its purposes in a pamphlet as promoting a general culture of internationalism to 'to prepare a generation which will carry over on to the international plane ideas of brotherhood, understanding, and mutual assistance'.[2] The work of UNESCO and the New Education Fellowship, which Tagore participated in, marked a continuation of these efforts to define and implement an idiom for international education.

Like Einstein, Tagore's interest in science and education had begun in childhood. In fact, a paternal uncle, Girindranath Tagore, had set up a scientific laboratory in the Jorasanko joint-family home, where chemistry experiments were conducted along with other experiments using battery-operated devices. In Rabindranath's memoirs, *Jivansmriti and Chelebela,* as well as in *Visaparichay* (a textbook on science that he wrote later in his life), he describes his education at Jorasanko, which included anatomy and scientific experiments. He also relates how a trip he made to the hill station Dalhousie with his father Debendranath, when he was twelve, stimulated his interest in astronomy and how these fascinations with astronomy and physiology led him to a large volume on astronomy by Sir Robert Ball and other volumes by Newcombe and Flamryon, as well as a set of articles on physiology by Huxley. Rathindranath Tagore mentions that when he and his father visited England in 1920, they made a special point of visiting the observatory in Greenwich, where the Astronomer Royal showed him the photographic plate of the solar eclipse that was said to confirm Einstein's theory of general relativity. Tagore dedicated his book on science *Our Universe* to Satyendranath Bose, whom Einstein had earlier recognized for his work on light quanta. His knowledge of quantum physics was reinforced by his interest in the findings of Niels Bohr and his meetings in India with physicists Arnold Sommerfeld in 1928 and Werner Heisenberg in 1929.

Both Tagore and Einstein had rebelled against conventional forms of schooling and argued for more flexible, creative forms of education. Of his early educational experience, Tagore says:

> The highest education is that which does not merely give us information but makes our life in harmony with all existence. But we find that this education of sympathy is not only systematically ignored in schools, but it is severely repressed. From our very childhood habits are formed and knowledge is imparted in such a manner that our life is weaned away from nature and our mind and the world are set in opposition from the beginning of our days. Thus the greatest of educations for which we came prepared is neglected, and we are made to lose our world to find a bagful of information instead. We rob the child of his earth to teach him geography, of language to teach him grammar.[3]

Rabindranath soon refused to attend school and received no diplomas. Instead, he devoted forty years of his life towards developing a creative learning centre in the natural setting of Santiniketan, which evolved into the centre of higher learning Visva-Bharati and an Institute at Sriniketan dedicated to rural education and incorporating scientific methods in aid of better health conditions and agricultural procedures.

Dmitri Marianoff reports that Einstein spoke of his hatred for the old mechanical methods of pedagogy in Munich, with its militaristic instructors and how everything was stilted and academic, the cramming of irrelevant information to pass examinations. When asked what he would teach, Einstein replied in a manner very reminiscent of Tagore's approach:

> Nothing that would overburden one side of the youthful mind. Why occupy our minds with so much unnecessary baggage, no part of which ever serves us in later life? Instead of this, why not teach natural science that will bring us more in contact with the world? From dead arts, they teach us dead things. That is not life, nor a serious part of life.[4]

Later, when he received an honorary degree from the State University of New York in Albany in 1936, he chose to speak on education, again on lines reminiscent of Tagore:

> To me the worst thing seems to be for a school principally to work with methods of fear, force and artificial authority. Such treatment destroys the sound sentiments, the sincerity, and the self-confidence of the pupil. It produces the submissive subject.
>
> . . . The point is to develop the childlike inclination for play and the childlike desire for recognition and to guide the child over to important fields for society. Such a school demands from the teacher that he be a kind of artist in his province.[5]

Regarding their personal meetings, which number at least six, there seems to be some controversy regarding precise dates and places, which may in part be due to the fact that the peripatetic Tagore travelled so extensively during his trips to Europe and met so many prominent people of that era. In a 1931 typed memoir that is kept in Rabindra Bhavana, Santiniketan, Tagore's writes concerning their meetings: 'I met Einstein during my first visit to Germany after the War. I was then deeply impressed by his great simplicity. There was nothing stiff about him-there was no intellectual aloofness. He seemed to me a man who valued human relationship and he showed toward me a real interest and understanding'. It is reported that their first meeting took place in 1926, when they met twice: once for a breakfast at the Prussian Ministry of Culture, and again at Einstein's home in Berlin. This meeting would have been after his visit to Italy in the spring of 1926, at the official invitation of Mussolini. Following that visit, he had spent time with Romain Rolland in Switzerland, where he also met Italian dissidents and learned of how his comments in Italy were being used by the fascist government. This resulted in Tagore publishing an open letter in the *Manchester Guardian* condemning the fascist regime. The 1926 Tagore-Einstein meeting resulted in a photograph, as well as a later note from Einstein to Tagore in which he offers to be of any assistance to the poet, but no record of their conversations. In the typed 1931 memoir that has been preserved in the Rabindra Bhavana archives, Santiniketan, Tagore writes of their 1926 meetings:

> I recall that we talked on that occasion of whether or not I believed in modern industrial improvements to help us in our modern life. I told him then, and I have no reason to change my opinion now, that they were essential to our physical well-being; and inasmuch as nothing could stop these improvements, we should seek to use wisely what man's ingenuity had created out of his necessity. For we had reached that degree of civilization when we could no longer scratch with our fingers; we were using our intelligence to overcome through machinery the weakness of our limbs. Both Einstein and I held similar opinion that it was necessary for us, by such means as mechanical invention offered, to make use of Nature's store-house.[6]

Their next meetings took place in 1930, and would have taken place during a very busy schedule of Tagore's, which included, among other activities, an exhibition of his paintings at Galerie Pigalle in Paris and Birmingham and the presentation of his Hibbert lectures in Oxford in late May. There is general agreement that the first of the 1930 meetings took place on July 14th in Einstein's villa at Kaputh, near Berlin. Their conversations were recorded by Amiya Chakravarty and conducted through an interpreter, Dimitri Marianoff, who later married Einstein's step-daughter Margot, and noted in his account that, 'Tagore came often to see Albert'. Tagore's recollection is as follows:

> When I again visited Germany in the summer of last year, I was invited to come to see Einstein, this time in his own home in Kaputh, a short distance from Berlin. The Professor had a very beautiful retreat there, built high up on a hill. He came down to the road to meet me. His shock of white hair, his burning eyes, his warm manner again impressed me with the human character of this man who dealt so abstractly with the laws of geometry and mathematics. Together we walked slowly up the hill, and on his balcony we had afternoon tea. Between us was a sympathy which only the barriers of language made awkward. His mastery of English is not enough for conversation and I cannot speak German. The interpreter between us must have had a not too easy task.[7]

As well as encompassing their interest in philosophy and science, the visit acknowledged a mutual love of nature. Tagore, of course, was one of the world's great poets of nature, and his educational scheme was predicated on learning in a natural setting. Marianoff notes that Einstein's could no more do without nature 'than his meat or drink—the imperative need of his walks in the out of doors. The vastness of the mountains, the new grass, the brilliant dew on green leaves, white birch trees, dark pines, broad fields, pasture lands, sounds, sights, smells of the earth, sunrise, sunset—the whole great living story. The sentences he reads here are stronger, finer, more splendid in their phrasing than any words he finds in books'.[8] That Einstein met Tagore and immediately took him into his garden, rather than the veranda or living room, was a gesture that he reserved for those he held in 'special affection'.[9] The Kaputh location would have been the setting for their first dialogue with Amiya Chakravarty and Marianoff in attendance. The dialogue, which was later entitled 'On the Nature of Reality', begins with Tagore acknowledging that while Einstein has been busy hunting down with mathematics the two ancient entities, 'Time and Space' he has been lecturing in Oxford on the 'eternal world of Man, the universe of Reality'.

A second visit took place about a month later on the 19th of August at the beautiful villa of Toni Mendel, a close friend of Einstein's, at Wannsee, a suburb of Berlin, where she lived with her daughter Herta and son-in-law, scientist Dr. Bruno Mendel.[10] Tagore had just visited Coblenz, where the German Youth Movement was developing, and the dialogue that took place on this occasion was later entitled 'Youth and Music'. He wrote in his memoir:

> Professor and I met again, at the home of Dr. Mendel, who is occupied with the investigation in the field of cancer. Once more I was able, in Einstein, to note the innate modesty of the man. He has a vitally acquisitive mind, and while I haven't much training to meet him on mathematical grounds, we had many discussions together which approached the boundary line of human and abstract conceptions of Reality. Einstein has often been called a lonely man. Insofar as mathematical symbols help liberate the mind from trivialities, I suppose he is a lonely man. His is what might be called a transcendental materialism, which reaches the frontier of metaphysics, where there can be utter detachment from the entanglement of the world of self. To me both Science and Arts are expressions of our spiritual nature where they are above our biological necessities, and possessed of an ultimate value.[11]

It is not surprising that their dialogue would focus on the development of youth and music, two subjects that were never far from their minds. Both were steeped in music from an early age, and it was integral to their personalities. Einstein played the piano and violin from an early age and was known to start his day playing classical music. Tagore, composer of over 2,000 songs, including the national anthems of India and Bangladesh, as well as inspiring the Sri Lanka anthem, was deeply rooted in the Indian classical tradition and familiar with various folk forms of music.

Two versions of the July 14th and August 19th dialogues were published, one carried by *The New York Times* on August 10th, following the first dialogue and edited by Einstein, and another slightly different version that Tagore edited and published along with his Hibbert Lectures.

After the August 19th visit with Einstein, Tagore spent a month in Switzerland, where he met with Romain Rolland, departing for Russia in mid-September with Margot Einstein, Dr. Harry Timbers, Amiya Chakravarty, Dmitri Marianoff and his grand-nephew Saumendranath, When he came back from Russia, sometime in late September, Tagore again met Einstein at the home of Toni Mendel, her daughter Hertha Mendel and son-in-law Dr. Bruno Mendel. There appears to have been no record of the conversation that took place during this visit, but no doubt it would have included an extended discussion on Tagore's impression of Russia, a subject that was of deep interest to both men.

Einstein around this time agreed to be one of the sponsors of 1931 *Golden Book of Tagore*, along with Mahatma Gandhi, Jagadish Chandra Bose, Romain Rolland and Kostis Palamas. At the end of his article, Einstein addressed Tagore:

> You saw the fierce strife of creatures, a strife that wells forth from need and dark desire. You saw the withdrawal in calm meditation and in creation of beauty. Cherishing these, you serve mankind all through a long and fruitful life, spreading everywhere a gentle and free thought in a manner such as the seers of your people have proclaimed as the ideal.[12]

Tagore sailed for New York in early October, where he met President Hoover and spoke to a standing-room-only audience at Carnegie Hall. Einstein and his wife left for American on 2 December, 1930. They spent December 10–15th in New York and during that period, Tagore and Einstein met for a final time, in which they spent a morning in 'animated conversation', producing a striking photo with the caption: 'A Mathematician and Mystic Meet in Manhattan'.[13] Einstein continued on to California and a research appointment at the California Institute of Technology that lasted several months before his return to Germany. Tagore left for India at the end of the month.

What brought them into contact in following years not so much their common interests in education, music or even science, but rather concerns over a developing world crisis that included the rise of Nazism and the persecution of Einstein. By December 1931, the Nazis under Hitler were gaining strength, and Einstein accepted a reappointment at the California Institute of Technology, noting in his travel diary that he had made his decision to give up his Berlin position. Within the next several years, the Nazis had raided and confiscated his house at Kaputh and the Berlin apartment, blocked his bank account and declared him an enemy of the Third Reich. Following the confiscation of his property, Einstein presented his passport to the German consulate in New York, saying he had 'no further use for it'.[14] In the fall of 1933, Einstein accepted an appointment at the Institute for Advanced Study, Princeton, where he continued to pursue his physics and mathematics and to continue his humanitarian efforts until his death in 1955.

Tagore expressed his concern over Hitler's regime and the treatment of Einstein in a 1934 letter to N.E.B. Ezra: '[If] the brutalities we read of are authentic, then no civilized conscience can allow compromise with them. The insults offered to my friend Einstein have shocked me to the point of torturing my faith in modern civilization'.[15] He was also said to have refused an honorary degree from Berlin University as a mark of protest against the persecution of Einstein in Nazi Germany. [16]

Notes

1 *Selected Letters of Rabindranath Tagore*, edited by Krishna Dutta and Andrew Robinson (Cambridge: Cambridge University Press, 1997).
2 WFEA (World Federation of Education Association), 1929: 94, quoted in Sylvester, Robert, 'Mapping international education', *Journal of Research in International Education* I(1) (09/2002), p. 104.
3 Tagore, Rabindranath, 'My school', *Personality* (London: Macmillan and Co., Ltd., 1917), pp. 116–17.
4 Marionoff, Dmitri, *Einstein: An Intimate Study of a Great Man* (New York: Doubleday, Doran and Col, Inc., 1944), pp. 34–5.
5 Pais, Abraham, *Einstein Lived Here* (New York: Oxford University Press, 1994), pp. 209–10.
6 Quoted in Singer, Wendy, 'Endless dawns' of imagination', *The Kenyon Review* 23(2) (Spring 2001), pp. 5–6.
7 Ibid.
8 Marianoff, *Einstein*, pp. 34–5.
9 Ibid.
10 For further details on the MendeRathindranathls, see Hoffman, Dieter, *Einstein's Berlin* (Baltimore: Johns Hopkins University Press, 2013), pp. 151–53.
11 Singer, 'Endless dawns' of imagination', p. 6.
12 *Golden Book of Tagore* (Calcutta: Golden Book Committee, 1931), p. 12.
13 *Selected Letters of Rabindranath Tagore*, 529.
14 See Marianoff, *Einstein*, pp. 146–47.
15 *Selected Letters of Rabindranath Tagore*, p. 437.
16 Ibid., p. 529.

1 Einstein, the reality of space and the action–reaction principle

Harvey R. Brown and Dennis Lehmkuhl

> For me it is an absurdity to ascribe physical properties to 'space'.
>
> Albert Einstein to Ernst Mach[1]

> That a real thing has to be presupposed as the cause for the preference of inertial systems over non-inertial systems is a fact that physicists have only come to understand in recent years.
>
> Einstein[2]

Introduction

It was striking in the 2012 seminar 'The Nature of Reality' at the Indian Institute of Advanced Study, Shimla, that so many speakers independently referred to the quip by Albert Einstein in his 1930 dialogue with Rabindranath Tagore that he, Einstein, was more religious than the great Bengali polymath. Of course, what Einstein was getting at is the fact that at the heart of the physical sciences, at least as he saw it, is a commitment to the reality of an external world whose goings-on are governed by laws that contain no fundamental reference to conscious agents, and in particular to human observers. For Tagore, all truth is human truth, if one is to take his claims literally.[3] For Einstein, the ultimate truth about the physical world transcends the human realm. Einstein was a 'realist', but his realism was of a modest, or non-metaphysical, kind. It is the job of physicists, he argued, to come up with models of a mind-independent reality that are explanatory of our phenomenological experiences of natural regularities, within the laboratory and without. However, whether such models, when judged successful, correspond to the way the world 'really' is, is a question Einstein thought best to leave aside. There are some philosophical questions for which Einstein thought the best response is a smile, and this was one of them. But Einstein stressed that even this weak, pragmatic take on truth involved a leap of faith. He made it clear, particularly in his 1949 *Autobiographical Notes*, that Nature's connivance in allowing for the success in the scientific venture as he conceived it could not be a foregone conclusion.[4] It was conceivable for Einstein that the mind, say, could be the bedrock of reality, but he felt that there was no good reason to start with that premise, and good reasons not to. Realism for Einstein was more a programme than a doctrine; it was a dogma about which he was careful not to be dogmatic.

Several years after the development of his 1915 general theory of relativity (GR), Einstein began to stress that physical space, or rather, the metric field, not only constitutes a fundamental, autonomous element of objective reality, it plays a causal role in accounting for the inertial motion of bodies.[5]

He compared this with the active role of space in the cases of Newtonian mechanics (NM) and special relativity (SR). In these cases, such putative action is clearly not reciprocated by bodies or fields: they do not act back on space-time structure, so the so-called action–reaction principle is violated. In contrast, in his relativistic theory of gravity (GR), Einstein was to see the vindication of the principle. The metric can have a dynamical life of its own in the absence of matter fields (though, as we shall see, this goes against Einstein's original expectations), but, more to the point, when the latter exist, the metric affects and is affected by them.[6]

Whether Einstein's reasoning is correct is open to doubt. What is debatable is not the claim that GR is consistent with the action–reaction principle; it is the claim that the older theories involving absolute space-time structures are not. More specifically, it is the assertion that such structures act in the relevant sense. Clearly, if they do not (as Newton himself argued regarding space; see below), then the fact that material bodies do not act back on them constitutes no violation of the action–reaction principle. The case for the view that absolute space-time structures of the kind that appear in NM and SR are not fundamental causal entities in their own right, but rather codifications of certain bare facts concerning the movement of bodies or behaviour of fields has been made a number of times, in different ways.[7] One curious instance is arguably in a letter Einstein himself wrote to Ernst Mach in late 1913;[8] another is due to Moritz Schlick in correspondence with Einstein in 1920.[9] We shall return to the second case below; it turns out to be directly related to the main concern of this chapter, which is, why would Einstein start systematically emphasising the action–reaction principle in his defence of GR only in the 1920s?

The action–reaction principle

It is a venerable tradition in natural philosophy to assert that a substance is the seat of actions on other substances, and in turn subject to the actions of these other substances: the action–reaction principle (AR). In his pre-*Principia* manuscript *De Gravitatione et Aequipondio Fluidorum*[10] (or *De Grav* for short), Newton insisted that natural philosophers tacitly understand substance as an entity that acts on things, even if they don't state this explicitly. He distinguished two kinds of action, one that stimulates the perceptions of thinking beings, and one between material bodies, as in collisions. (Later, he would extend this second kind to action at a distance.)[11] This distinction today seems of little import, and the fundamental premise, that it is interactions between systems that count in physics and not their intrinsic properties, is close to the heart of the 'structural realist' position that has been much discussed in recent years in the philosophy of science. Be that as it may, Newton is clearly appealing to a principle in the *De Grav* that is more fundamental and general than what he would later designate as his third law of motion in the *Principia*—though the latter is often referred to as the law of action–reaction. (We shall see shortly how space, for Newton, is a kind of exception to this fundamental principle.)

Leibniz, whose views on the nature of space and time were so different from Newton's, nonetheless shared the same intuition. In fact, when defining substance as that which acts and can be acted upon, he understood that he was adopting the view of the scholastics.[12] Caution must be exercised, however, in attributing AR, as it is standardly understood today, to Leibniz. In the light of his doctrine of pre-established harmony, the meaning of causation, or rather interaction, within his deep metaphysics is almost certainly at odds with those views adopted by the majority of current metaphysicians,[13] not to mention physicists—and its scope is still controversial.[14]

If there is a questionable aspect of AR, it is less the claim that substances act (how otherwise could their existence be known to us?) than the notion that they are necessarily acted back upon, that action must be reciprocal. If all substances act, they do so in relation to other substances; these other substances therefore cannot be immune from external influences. Now it might seem arbitrary on *a priori* grounds to imagine that the 'sensitivity' of such substances is not universal. That is to say, it might seem arbitrary to suppose that not all substances react to others. But no such abstract qualms can be entirely compelling; Nature must have the last say.

Nowadays, it is well known in the foundations of physics that the de Broglie-Bohm ('pilot wave') interpretation of quantum mechanics, in its standard form, involves a dynamical agent (the wave function) that acts on corpuscles (point particles) but is not acted back upon, at least by the corpuscles. It is noteworthy that David Bohm himself clearly found such violation of AR uncomfortable in his (first) 1952 formulation of the theory and attempted to remedy it, as have others after him.[15] While defenders of the original pilot-wave theory can legitimately argue that AR is not a logical necessity, others see its violation as a blemish in the theory.[16] It seems fair to say that currently within the physics community, there is no consensus supporting the failure of AR in quantum theory.

For his part, Einstein himself had already stated in 1922 that it is 'contrary to the mode of scientific thinking to conceive of a thing . . . which acts itself, but which cannot be acted upon'.[17] The object of Einstein's ire in 1922 was NM and his own creation, SR. Yet there is no hint in his writings around the time of the development of SR in 1905 that Einstein considered either of these theories to incorporate a violation of the action–reaction principle; at any rate, the explicit condemnation came later. Why? In all probability, because it was part of an honest sales pitch for GR, his greatest and most radical contribution to science, after Einstein was reluctantly forced to concede, because of results by de Sitter, that the theory as a whole was not consistent with 'Mach's Principle', even though special solutions are. It seems that this change of tack on Einstein's part was consolidated in the mentioned 1920 correspondence with the physicist-philosopher Moritz Schlick.

But before examining the evolution of Einstein's views on the causal role of space-time, it is worth briefly visiting Newton's own views on the nature of absolute space, especially as expounded in the *De Grav*. This manuscript contains a hard-hitting critique of Descartes's relational theory of motion and the positive reasons why Newton felt compelled to posit the existence of absolute space; these important lines of reasoning have been discussed extensively in the literature.[18] What concerns us is Newton's insistence that space, despite its reality, does not act on either our sense organs (which is patent) or on our bodies. If it is a substance, it is by Newton's own reasoning *sui generis* in its causal inefficacy, and ultimately, Newton had no interest in classifying it as either substance or accident.[19]

When, in the context of discussions relevant to this chapter, Newtonian space is assigned a causal role, it is usually to account for inertia, i.e., the privileged existence of inertial frames, or equivalently, the special motions of force-free bodies. In the *De Grav*, Newton explicitly renounced such a role. He stated that the reason why projectiles that are not being acted upon by other bodies moving in straight lines and at uniform speed is precisely that space has no ability to help or hinder any change in the motion of bodies![20]

Einstein on absolute space

Einstein and Mach

Einstein's tortuous road to the 1915 field equations of his unhappily named general theory of relativity followed a number of fundamental, partially overlapping, conceptual signposts, apart from the requirement of securing the Newtonian limit in the case of suitably defined weak gravitational fields. These were the so-called equivalence principles, which connect gravitational effects with inertia, Mach's ideas concerning the origins of inertia, the principle of the relativity of motion and finally, the conservation of energy-momentum. It is probably no exaggeration to say that none of these guiding principles was to survive fully intact once the promised land was reached. In this section, we shall be concerned in particular with Mach's influence on Einstein's thinking, because its demise is intimately connected with the appearance of AR in Einstein's arsenal of arguments in favour of GR. The story of Einstein's debt to Mach has of course received considerable attention in the literature;[21] we shall deal only with those elements that are necessary for our ends.

Mach

Ernst Mach's own thinking on inertia is usefully divided into strands: his critique of Newtonian mechanics, more specifically, of absolute space and time, and his somewhat vague and varied suggestions as to how to remedy the theory. As for the first strand, Mach's central objection to NM is that in appealing to motion with respect to absolute space in accounting, say, for centrifugal effects in rotating bodies, one is being unfaithful to the fundamental aim of science, which has to do with providing an economic systematisation of experience. Space is not observable, so a purportedly fundamental epistemological (or perhaps better methodological) principle requires us, according to Mach, to refer only to motion relative to other bodies. It is not entirely clear that Mach sees space as 'acting' on a rotating body within the standard Newtonian account, and anyway, there appears to be no hint of a complaint in Mach's writings that it violates the action–reaction principle. The issue is the methodological/epistemological role of observables in physics associated with Mach's brand of empiricism, not the metaphysics of action.

Further insight into Mach's thinking is seen in the second, positive strand. Opinion is divided as to whether he was suggesting a reformulation of Newtonian mechanics or a new, distinct theory accounting for the inertial motion of bodies. John Norton, for example, is sceptical about the seriousness of Mach's revolutionary intentions;[22] Julian Barbour is not. Barbour indeed provides powerful textual evidence that Mach, on at least one occasion, was searching for an explanation of the inertia (as opposed to inertial mass) of a force-free body, i.e., its uniform, rectilinear motion, in terms of distant bodies, in analogy with the explanation of the acceleration of a body resulting from the gravitational influence of distant masses—an account that would yield the Newtonian predictions to a good approximation in the case of a universe populated to the extent ours is.[23]

There are three points to be made at this stage. First, if Mach's concern were (counterfactually) truly AR, would it be beyond the bounds of reasonableness to suppose that he might have sought to make space dynamical and capable of being acted on by bodies? There is no hint of this possibility in Mach's writings, despite the fact that dynamical notions of space had already been proposed in 1872 by Zöllner and independently in 1876 by Clifford.[24]

Mach in fact showed considerable interest in non-Euclidean geometry and thought it probable, but not inevitable, that it would have no role to play in physics. If it did, it would for Mach only be through the behaviour of observable, material bodies, and there was no evidence for it when he wrote essays for *The Monist*, which would be published collectively as *Space and Geometry* in 1906. He thought that the behaviour of matter was as unlikely to indicate the reality of non-Euclidean geometry as it would satisfy 'the atomistic fantasies of the physicist'.[25]

Second, it would seem that Mach's gravitation-like proposal for the origin of inertia and the very existence of inertial frames would involve action at a distance.[26] In fact, it could be called super-action-at-a-distance; the inertial motion of a body is being attributed to the existence of celestial bodies *so far* away that their gravitational actions on it are negligible. The caveat is that Mach's own notion of causality was rather thin; Norton has called it 'idiosyncratic'.[27] Mach saw physics as providing only functional dependencies between experiences, systematic correlations rather than causal interactions (insofar as the distinction is meaningful). The commitment to the notion that '*the law of causality is sufficiently characterised by saying that it is the supposition of the mutual dependence of phenomena*' on Mach's part[28] perhaps explains why he was comfortable enough with Newton's picture of gravity to encourage the search of an analogous account of inertia. However all this may be, it seems reasonable to conclude that Mach was not obviously concerned, in the context of his analysis of inertia in NM, with the action–reaction principle or its violation, in anything like the ordinary sense. (It is true that Mach provided an operational reading of inertial mass based on Newton's third law of motion, one that proved to be influential. But as we have urged, the third law is not to be conflated with the AR principle.) In fact, a degree of resonance is discernible between Mach's view of the nature of causal connections and that of Leibniz, despite the chasm between their views on metaphysics.

The third point is closely related to the second. An important aspect of Mach's thinking about inertia is the emphasis on the cosmological nature of its origins. It is not some subset of distant bodies that determines the system of inertial frames: It is the totality of bodies in the universe. Barbour has aptly connected this cosmological strand of Mach's reasoning—the requirement of *self-referentiality* in any adequate account of the observed world—with Kepler's 1609 theory of place and motion.[29] Prior to 1917, the strand played only a minor role in Einstein's thinking.

Back to Einstein

Einstein's decade-long love affair with Mach's philosophy of inertia was complicated and tortuous. The first complication is that Einstein entertained two quite distinct Mach-inspired doctrines, one of which actually had no basis in Mach's writings, as Barbour first emphasised in 1990.[30] This was the doctrine that the *inertial mass* of a body is to be explained as arising from the presence of other bodies, with the consequence that a body at spatial infinity should have zero mass. (Mach himself had no difficulty in viewing inertial mass as an intrinsic property of the body, and, as mentioned above, used Newton's third law to reveal its operational significance.) Indeed, it was this idea of the 'relativity of inertia' that Einstein had in mind in his first endorsement of Mach's reasoning in a paper published in 1912.[31]

It has to be noted that this paper was written at a remarkable moment of time. In March 1912, Einstein had completed his final paper on a theory of static gravitational fields in which gravity was represented by a scalar *c*, which was supposed to represent not only the gravitational potential at the space-time point in question, but also the speed of light at that very space-time point.[32] In August 1912, Einstein moved from Prague to Zurich, where his collaboration with Marcel Grossmann began, one which culminated in the first paper[33] propounding what is known as the *Entwurf* theory, in which gravity is represented by a dynamical ten-component metric tensor field $g_{\mu\nu}$ whose non-dynamical counterpart Einstein already knew from Minkowski's 1908 formulation of special relativity. Thus, the cited 1912 paper was written only a few months before Einstein learned about the powers of the metric tensor, and it is noteworthy that he had already endorsed the idea of a relativity of inertia in the context of a relativistic theory of gravity.

Einstein states in the 1912 paper[34] that the results he has obtained give support to the (alleged) idea of Mach that the inertia of point masses is a result of the presence of other masses, that it rested on an *interaction* (*Wechselwirkung*) of the point particle with those other masses. (Recall that Mach himself did not use the term 'interaction' but spoke of mutual dependencies.)[35] We find similar statements in the mentioned 1913 paper with Grossmann and another 1913 paper by Einstein;[36] in each instance, Einstein emphasises that the inertia of a body should be derived as the result of an interaction of this body with other bodies.[37] Also in 1913, Einstein wrote two letters to Mach, the first in June, the second in December.[38] In the June letter, Einstein writes enthusiastically that in his *Entwurf* theory that he developed with Marcel Grossmann, '*inertia* has its origin in some kind of *interaction* of the bodies, completely in the sense of your considerations about Newton's bucket experiment'. Despite the wording, careful reading of the letter makes it clear that again Einstein is thinking of inertial mass rather than the inertial frame.[39]

However, in a 1914 paper,[40] Einstein explicitly discussed Newton's famous bucket thought experiment and Mach's criticism thereof—arguing that if we have a choice, we should go for an account in which no absolute motion exists. Here, though not for the first time, Einstein is advocating something much closer to Mach's own concerns, and it was to dominate his thinking for at least the next two years, as we shall see. At this point, it is worth noting that in his well-known 1999 study of the principle of general covariance, John Norton argued that as early as 1913, Einstein articulated a connection between Mach's theory of inertia as he understood it and the action–reaction principle. Indeed, Norton sees concern with the violation of AR in pre-GR space-time theories as 'the enduring core of the cluster of ideas that led Einstein to the relativity of inertia and Mach's principle'.[41] We shall defer discussion of this claim to the section, 'Whence AR?'.

Hope of a solution?

How was the Machian positive programme related primarily to inertial motion (rather than to inertial mass) to be implemented? Essentially, by finding a way to bypass it! Einstein was to link the problem of inertial motion with a notion he expressed clearly in 1911, itself related to the equivalence principle: that a uniformly accelerating reference frame (which reproduces all the effects of a homogenous gravitational field) is no more absolute than an inertial frame. Indeed, he hoped that his future theory of gravity would allow for a yet further generalisation of this putative extension of the Galilean-Einstein relativity principle—to *all* frames, such that the very distinction between inertial and non-inertial motion would become relative, non-absolute.[42] By 1912, Einstein was convinced that the success of the complete 'relativity of motion' would be guaranteed if the gravitational field equations turned out to be generally covariant.[43] What is important for our purposes is that Einstein saw relativity of inertia, the principle of the relativity of motion and the equivalence principle as walking hand in hand. As Barbour has stressed:

> The drift of Einstein's thought is now clear. Whereas the logic of Mach's comments called for explicit derivation of the distinguished local frames of reference from a relational law of the cosmos as a whole, Einstein is working towards elimination of the problem of the distinguished frames by asserting that they are not really distinguished at all.[44]

Renn and Sauer have also emphasised essentially the same point (though with less emphasis on the departure from Mach's cosmological speculations):

> Einstein's view that it made sense to search for a generalization of the relativity principle of classical mechanics and special relativity was . . . based on his acceptance of a philosophical critique of classical mechanics raised by Mach and others. According to this critique, the justification of the privileged role of inertial frames of reference by the notion of absolute space was problematic, while the inertial forces experienced in accelerated frames of reference require an explanation in terms of the interaction between physical masses. Such an explanation would then eliminate any need for absolute space as a causal agent in the analysis of motion. The generalized relativity principle would go, so at least was Einstein's expectation, a long way, and might actually go all the way, towards an implementation of Mach's critique of classical mechanics in the new theory of gravitation.[45]

The culmination of this reasoning would appear in Einstein's great review paper of 1916.

1916

At the start of his 1916 review paper on the new general theory of relativity,[46] Einstein would provide several (too many?) motivations for the requirement of general covariance; one of them was Mach's analysis of inertia. He went on to introduce a famous thought experiment involving two fluid (i.e. not entirely rigid) bodies S_1 and S_2 in empty space rotating around each other at constant angular velocity and so far apart and isolated that the gravitational force between them and all other bodies is negligible. Suppose it is a verifiable fact that body S_1 is spherical in shape, while body S_2 is ellipsoidal. Einstein asserts that, in answering the question as to why this is so,

> No answer can be admitted as epistemologically satisfactory, unless the reason given is an *observable fact of experience*. The law of causality has not the significance of a statement as to the world of experience, except when *observable facts* ultimately appear as causes and effects.
>
> Newtonian mechanics does not give a satisfactory answer to this question. It pronounces as follows: The laws of mechanics apply to a space R_1, in respect of which the body S_1 is at rest, but not to a space R_2, in respect to which the body S_2 is at rest. But the privileged space R_1 of Galileo, thus introduced, is merely a *factitious* cause [*fingierte Ursache*], and not a thing that can be observed. It is therefore clear that Newton's mechanics does not really satisfy the requirement of causality in the case under consideration, but only apparently does so, since it makes the factitious cause R_1 responsible for the observable [shape] difference in the bodies S_1 and S_2.

It is clear in this passage that despite Einstein's talk about the 'law of causality'—for which Mach's understanding would probably be very different from Einstein's—his critique of NM is very similar to Mach's, and it is explicitly described as epistemological in nature. Indeed, the 'epistemological' shortcoming of NM had already been mentioned by Einstein in 1913.[47] But a notable feature of Einstein's Machianism is its selectivity. Mach rejected the existence of atoms on essentially the same grounds that he rejected the existence of absolute space; Einstein did not. By 1905, Einstein had made important contributions to the theory of capillarity, molecular dimensions and statistical mechanics—not to mention the nascent quantum theory. His 1905 paper on Brownian motion in particular would provide the basis for arguably the first significant empirical evidence of the existence of atoms. As Don Howard has written:

> A careful reading, especially of Einstein's papers on the foundations of statistical physics, reveals that the influence of Mach and Ostwald was being felt. It was not, however, in the form of any doubt about the reality of atoms, but in the form of a caution about prematurely investing these atoms with any properties other than those necessary for the purpose at hand. . . . It was thus not an ontology of unobservables that troubled Einstein; it was merely an ontology that was richer than it need be.[48]

But the case of immaterial space was different. In his 1916 paper, Einstein went on to claim that the proper cause of the difference between the bodies S_1 and S_2 must be sought in the relations between S_1, S_2 and other physical bodies, rather than in their relations to unobservable absolute space.

> These distant masses (and their motions relative to S_1 and S_2) must then be regarded as the seat of the causes (which must be susceptible to observation) of the different behaviour of our two bodies S_1 and S_2. They take over the role of the factitious cause R_1. Of all imaginable spaces R_1, R_2, etc., in any kind of motion relatively to one another, there is none which we may look upon as privileged *a priori* without reviving the above- mentioned epistemological objection. *The laws of physics must be of such a nature that they apply to systems of reference in any kind of motion.* Along this road we arrive at an extension of the relativity principle.[49]

The connection in Einstein's thinking between the problem of inertia and the generalised relativity principle is clearly stated here. In the previous year, he had regained confidence in the general covariance of the gravitational field equations (following the notorious struggle with his 'hole argument') and thus found a purported solution to the Machian problem simply on the basis of the fact that the 1915 field equations are generally covariant. It is ironic, then, that for most of the 1916 paper, Einstein reverts to the use of unimodular coordinates (for which $\sqrt{-g} = 1$, g being the determinant of the metric field $g_{\mu\nu}$) with the explicit purpose of simplifying the equations and calculations.[50] Be that as it may, it is worth noting, as Barbour did in 1990,[51] that insofar as general covariance was to play a role in solving the Machian conundrum, it was the relativity of motion, and not the relativity of inertial mass, that was relevant.

The year 1916 also marked the year of Mach's death. Einstein wrote an obituary, in which he discussed in detail Mach's criticism of Newton's concept of absolute motion, which in turn rested on the notion of absolute space. Einstein wholeheartedly subscribed to Mach's point that physics should avoid using these absolute concepts, and even makes fun of the 'philosophers' with their 'treasure chests of the "absolute" and the "a priori"'.[52]

Mach in trouble: 1917 to 1921

Yet Einstein in 1916 had not yet relinquished his 1912 interpretation of Mach's philosophy in terms of inertial mass. Einstein and Grossmann had already derived a related effect in the *Entwurf* theory,[53] namely that a mass point's inertial mass grows if other bodies with inertial mass are close by. In his 1921 Princeton lectures,[54] after de Sitter's attack on Mach's principle as defined in 1918 (see below), Einstein considered the limit where the geodesic equation goes over to the Newtonian equation of a mass point subject to a gravitational field, where $g_{\mu\nu} = \eta_{\mu\nu} + \gamma_{\mu\nu}$ and γ_{00} corresponds to the Newtonian gravitational potential assumed to be small to first order. He derived a form for the equations of motion that he interpreted as showing that, in GR too, the inertial mass of a mass point increases if other massive bodies are close by. This result he took as 'strong support for Mach's ideas as to the relativity of all inertial actions'. He went even further:

> If we think these ideas consistently through to the end we must expect the *whole* inertia, that is, the *whole* $g_{\mu\nu}$-field, to be determined by the matter of the universe [. . .]

Einstein then showed that this can indeed be achieved for the (static, spatially closed) Einstein universe.

Of course, this was already a fallback position. Originally, Einstein had expected that in *every* solution of the Einstein equations, i.e., in every possible universe, the mass-energy of matter $T_{\mu\nu}$ would uniquely determine the $g_{\mu\nu}$-field. De Sitter brought an end to this idea, as will be described below.

Either way, taking the (partial) relativity of inertial mass of a body as reason to believe that the $g_{\mu\nu}$-field should be entirely determined by the $T_{\mu\nu}$-field is a remarkable conceptual jump. But this notion of $g_{\mu\nu}$ supervening on $T_{\mu\nu}$ was far from new in Einstein's Mach-inspired thinking; it can be traced back to the *Entwurf* theory. Later, in 1918, it was taken by Einstein to *define* 'Mach's Principle'. What, then, had happened to his notion of linking Mach's philosophy with general covariance?

The issue had become much more complicated. Earlier in 1918, Kretschmann would famously question whether general covariance had any empirical content at all.[55] Kretschmann's analysis forced Einstein to clearly distinguish the relativity of motion principle from both the equivalence principle and from the view that the gravito-inertial field represented by the metric tensor must be determined exclusively by the distribution of masses in the universe.

In 1918, his precise formulation of 'Mach's Principle' became the claim that the $g_{\mu\nu}$ field must be 'conditioned and determined' by the mass-energy momentum $T_{\mu\nu}$ of matter.[56]

Alas, although no one seems to have appreciated it at the time, $T_{\mu\nu}$ cannot be defined independently of $g_{\mu\nu}$.[57] Einstein himself recognised this point explicitly in 1954, the year before his death; in a letter to Felix Pirani, he admitted that Mach's Principle was hopeless for exactly this reason.[58] However, even in 1918, other grounds for doubting the validity of the principle already existed. A solution of Einstein's 1915 field equations is a static metric field corresponding to flat, matter-free Minkowski space-time; this was one reason why in 1917 Einstein felt he had to introduce modified field equations with a cosmological constant to rescue the principle.[59] Alas, again, de Sitter would soon show that the modified field equations admit other vacuum solutions: the degrees of freedom of the metric field are in general not uniquely determined by the mass-energy of matter.[60] In the years between 1918 and 1922, Einstein was thus forced to admit the metric field $g_{\mu\nu}$ as a dynamical player in its own right according to GR, akin to the electromagnetic field $F_{\mu\nu}$.

Nonetheless, he still clung to the weak form of Mach's Principle present in the Princeton lectures, which holds only in the Einstein universe—at the time supposed to be the best model of the actual universe.

These lectures are also the place where Einstein for the first time published an explicit, clear connection between the Machian critique of absolute space and the action–reaction principle. Here, he stated:

> [I]t is contrary to the mode of thinking in science to conceive of a thing (the space-time continuum) which acts itself but which cannot be acted upon. This is the reason why E. Mach was led to make the attempt to eliminate space as an active cause in the system of mechanics [. . .] In order to develop this idea within the limits of the modern theory of action through a medium, the properties of the space-time continuum which determine inertia must be regarded as field properties of space, analogous to the electromagnetic field.[61]

This is hardly a faithful account of Mach's reasoning. More to the point, we witness here what seems to be an important shift in Einstein's thinking about the nature of the problematic role space(time) plays in Newtonian mechanics and special relativity. But the seeds for this shift were planted at least two years earlier, as we shall now see.

The Einstein-Schlick correspondence

The background

The growing recognition, on Einstein's part, of the tension between the field equations in GR and his 1918 version of Mach's Principle led him, as we have seen, to effectively assign genuine degrees of freedom to the metric field in the general case (not for the Einstein universe). This development finds a clear expression in a 1920 paper[62] where Einstein speaks of the electromagnetic and the gravitational 'ether' of GR as in principle different from the ether conceptions of Newton, Hertz and Lorentz. The new, generally relativistic or 'Machian ether', Einstein says, differs from its predecessors in that it interacts (*bedingt und wird bedingt*) both with matter and with the state of the ether at neighbouring points.[63] There can be little doubt that the discovery of the partial dynamical autonomy of the metric field was an unwelcome surprise for Einstein, that as a devotee of Mach, he had been reluctant to accept that the metric field was not, in the end, 'conditioned and determined' by the mass-energy momentum $T_{\mu\nu}$ of matter.

But by the time he gave his Princeton lectures in 1921, Einstein had clearly made virtue out of necessity. He had come to see the *inter* active nature of the metric field as a blessing in disguise. This transition in Einstein's thinking was first brought out in 1920, in the course of a correspondence Einstein had with Moritz Schlick.

Schlick started out as a physicist and graduated with a PhD in optics in 1904 under the supervision of Max Planck—maybe the only physicist that Einstein revered as much as Hendrik Lorentz. Schlick then turned to philosophy and was one of the first philosophers to analyse general relativity; his earliest article stems from 1915.[64] In 1917, Schlick wrote another paper for the widely read journal *Die Naturwissenschaften*. The article met with so much interest that it was commissioned to be reprinted as a monograph within the same year, with an enlarged and improved second edition being published in 1919. This was, of course, the year of Eddington's solar eclipse observations and Einstein's rise to international fame, so that another, again enlarged, edition became necessary in 1920.

In Einstein's first letter to Schlick, he was full of praise of Schlick's 1915 article on the relativity theory. He writes:

> Yesterday I received your paper and I already read it through completely. It is among the best that has been written on relativity to date. From the philosophical perspective, nothing nearly as clear seems to have been written on the subject.[65]

This is the first of at least 26 letters exchanged by Einstein and Schlick between 1917 and the end of 1920 alone. The frequency of correspondence reached its height in 1920, when Schlick was preparing the fourth edition of his book and when Einstein was struggling with the blow that de Sitter had dealt to Mach's Principle. An important point in the correspondence had to do with a footnote that was present in the first three editions of Schlick's book; it dealt with the way Einstein had criticised the absolute space of Newtonian mechanics in the thought experiment with the two fluid bodies we encountered in the last section. Schlick writes:

> One does not have to understand Newtonian theory as taking Galilean space—after all, an unobservable thing—for the *cause* of the centrifugal forces. Instead, one might take the talk of absolute space a mere restatement of the bare fact that these forces exist [. . .] One does not need to regard absolute rotation as the cause of the ellipsoidal shape [of S_2]. Instead one can say: the former is *defined* by the latter.[66]

Einstein comments on this remark in a letter to Schlick dated 21st March, 1917. He writes:

> Your criticism on page 178 (footnote) is legitimate. On close examination, the causality requirement is just not sharply defined. Various grades exist for satisfying the causality requirement. It can only be said

> that the general th[eory] of r[elativity] has been more successful than classical mechanics in satisfying it. It might be a rewarding task for an epistemologist to think this through carefully.[67]

Schlick may well have seen this as a challenge posed by Einstein to himself. At any rate, he did think the matter through carefully, and in 1920, while working towards the fourth edition of his book, Schlick published a paper entitled, 'Natural-philosophical considerations of the principle of causality'.[68] This paper sparked a new round of discussions between Schlick and Einstein, especially on the question of the differences between NM and GR.[69]

Action–reaction to the rescue

On the 5th June, 1920, Schlick sent Einstein the paper on the principle of causality, only four days before it was printed in *Die Naturwissenschaften*.[70] Einstein answered two days later with detailed comments on the manuscript. One of the issues he focused on is the question of whether the law of inertia has a different causal status in NM as compared to GR:

> Furthermore, on the question of the violation of the postulate of causality through the law of inertia. In your little book you rightly emphasised that I went too far in my exposition [of the 1916 thought experiment]. However, I cannot agree with your current position of the state of affairs. I think it would be correct to say: Newtonian physics has to attribute objective reality to acceleration, independently of the coordinate system. This is only possible if one regards absolute space (i.e., the ether) as something real. Newton does this in a coherent way. It does not matter whether you call that which you have to refer to in order to give reality to acceleration, absolute space, ether or preferred coordinate system. What remains unsatisfactory is the circumstance that this something enters *only one way* (*nur einseitig*) into the causal chain [. . .] The absolute space of Newton is independent, cannot be influenced, the $g_{\mu\nu}$-field of general relativity is subject to laws of nature, determined by matter (not only determining).[71]

Schlick answers within three days. Interestingly, he does not comment at all on Einstein invoking the action–reaction principle, but reiterates the point he had made in the above-mentioned footnote to his book, claiming that it is compatible with everything Einstein had written in his previous letter:

> . . . absolute space, which Newtonian mechanics of course has to presuppose, does not have to be considered by the latter as a cause in the sense of the principle of causality. In other words: Newtonian Mechanics does not have to consider inertial resistance in the context of certain

> kinds of motion as an effect of an absolute acceleration. It can instead take the former as the defining criterion of the latter.[72]

So while Einstein takes it as a given that absolute space acts in NM while spacetime *inter*acts in GR, Schlick questions the first point. Unfortunately, just as Schlick did not react to Einstein's new move of invoking the action–reaction principle, Einstein in his answer to Schlick on 30th June, 1920, does not react to this interesting position either. Instead, he offers an argument of why space has to be considered as absolute in NM:

> If I consider the equation
>
> mass · acceleration = force
>
> then "force" is something absolute (independent of the coordinate system), just like mass, if only the units (including the unit of length) are fixed. Hence, one also has to attribute an absolute status to acceleration. The latter is, with length and time, defined via d^2x/dt^2; thus, one must not also define acceleration via the law of inertia. Instead, one will have to choose to also consider x and t as absolute, i.e. physically sensible, quantities. For t this works out, if one does not take into account the difficulty of simultaneity c = practically ∞ via a clock; but for x this does not work out. One is forced to attribute a mystical, i.e. empirically inaccessible reality to space.[73]

We leave the reader to ponder over the validity and relevance of this strange argument, which appears to suggest a tension between the Galilean coordinate transformations, or rather the non-invariant transformations for the spatial coordinates, and the absolute nature of the acceleration caused by Newtonian forces. More pertinently, Einstein continues by pointing out how things fare better in GR:

> By the way, physical space possesses reality according to the general theory of relativity, too, but not an independent one; for its properties are completely determined by matter. Space is incorporated into the causal nexus without playing a one-sided role in the causal chain.

The second half of the first sentence is also striking, as Einstein had previously recognised that Mach's principle only holds for certain solutions of the Einstein field equations, not for all of them—but of course, at the time he considered those solutions for which it held as the only physically relevant ones. At any rate, we here see the complete position which would first be presented in the 1921 Princeton lectures: In Newtonian mechanics, space acts without being acted upon, while in general relativity, it *inter* acts.

In Schlick's answer of 29th August, 1920 (i.e., two months later), he succumbs to Einstein.[74] Even though Einstein had not addressed Schlick's alternative position of the status of absolute space in Newtonian mechanics, and Schlick himself failed to explicitly acknowledge the action–reaction principle as the decisive criterion, Schlick announced he was completely convinced by Einstein's argument and even points out that he would delete the critical footnote from further editions of his book.

The observability of the metric field

A noteworthy episode in the Einstein-Schlick correspondence was spurred by a remark in Schlick's 1920 *Die Naturwissenschaften* paper, that from 'an epistemological point of view it is remarkable that the gravitational field does not represent something observable in the same sense as the relative motion of visible bodies'. Einstein replied, in a letter dated 7 June of the same year:[75]

> It seems to me unjustified to state that gravitational fields should not be regarded as observable in the same sense as masses . . .

In turn, Schlick partially concedes the point, but there is a sting in the tail of his reply:

> I was probably not right with the assertion that a gravitational field was not observable in the same sense as masses. This does apply, at most, in the very rough sense that one may say: I do perceive two objects but not the gravitational field between them. However, it seems to me a debatable point whether in examining Machian thoughts the word 'perceptible' may be taken in the roughest sense.[76]

Schlick is correctly wondering what kind of 'observability' criterion is operative in Mach's critique of Newtonian absolute space. If it is the 'rough' one having to do with direct perceivability, then it is unclear how space-time structure in GR is to avoid Machian condemnation. On the other hand, suppose the relevant notion of observability is a more subtle one, that the gravitational field in GR is said to be observable because, like the electromagnetic field in the case of charged bodies, it has an effect on more directly perceivable masses. In this case, it is hard to avoid the conclusion that Newtonian space, insofar as it purports to explain inertia, is also observable, and the Machian condemnation of it collapses. It seems that in his correspondence with Schlick, Einstein is adopting something like the second position, abandoning the epistemological flavour of Mach's original critique of absolute space in favour of another tack: that based on the action–reaction principle. If this is correct, it is another example of the plasticity of Einstein's epistemological reasoning in the light of new developments in the physics (in this case, prompted by de Sitter's results), and it anticipates Einstein's well-documented turn to a more realist philosophy and a weakening of his empiricism.[77]

Whence AR?

As we mentioned above, Norton argued in 1999 that AR was in the back of Einstein's mind well before 1920, and indeed, formed the stimulus of his original Machian tendencies. Here is a further quote from Norton's study:

> This view of the deficiency of earlier theories [their violating the action–reaction principle] and general relativity's achievement is not one that grew in the wake of Einstein's disenchantment with Mach's principle. Rather, it was present even in his earliest writings beneath the concerns for the relative motion of bodies and the observability of causes.[78]

Indeed, this view is consistent with Einstein's 1922 claim, in the quotation found at the end of the section 'Hope of a Solution?' above, which (questionably) assigns AR as the true source of Mach's own misgivings about the role of absolute space in Newtonian mechanics. For his part, Norton quotes from one of the 1913 Einstein papers referred to in the Back to Einstein section above, the relevant passage being:

> It is a priori to be expected, even though it is not strictly necessary, that inertial resistance is nothing else but resistance against relative acceleration of the body *A* in question with respect to the totality of all other bodies *B*, *C* etc. It is well-known that E. Mach, in his history of mechanics, was the first who defended this position with sharpness and clarity, so that I can only refer to his work [. . .] I will call this position the "hypothesis of the relativity of inertia". [. . .] [A] theory in which the relativity of inertia is realised is more satisfactory than the current theory, for in the latter the inertial system is introduced; its state of motion, on the one hand, is not conditioned by the status of observable objects (and therefore caused by nothing accessible to perception) but, on the other hand, it is supposed to determine the behaviour of material points.[79]

But is this an unambiguous enunciation of the AR principle? Note the emphasis, once again, on the epistemological issue of observability and perception, which seems to foreshadow the way Einstein criticises NM in his 1916 review article. Furthermore, it is puzzling that there does not seem to be any other explicit commitment to the AR principle in Einstein's pre-1920 writings, including his effusive letters to Mach.[80] And as for Einstein's own 1921 interpretation of Mach, this could plausibly be attributed to a mixture of ignorance and opportunism. Finally, and perhaps most importantly, it seems to us that if there was a *single* core belief (of course, there could have been more) underpinning Einstein's adoption of the Machian relativity of inertia, it was probably the equivalence principle, and not AR.[81]

In his letters leading up to the paper in which the above quote appears, Einstein keeps emphasising that the fundamental idea on which everything else rests is that gravitational fields and 'acceleration fields'[82] are equivalent. Suppose the familiar notion is accepted that gravitation is some kind of interaction between massive bodies, despite the fact that Einstein is embarking on a radically non-Newtonian theory of gravity. Then, if gravitation and acceleration/inertia are to be truly equivalent, inertia *must* also be an interaction between massive bodies. Indeed, Einstein seems to incorporate such logic in his letter to Mach from 25th June, 1913:

> Next year the eclipse is supposed to show whether light rays are bent by the sun, whether, in other words, the fundamental assumption of the equivalence between acceleration of the frame of reference on the one hand and the gravitational field on the other is indeed right. If it is, then your brilliant investigations on the foundations of mechanics receive—despite Planck's unjustified criticism—a shining confirmation. For it follows with necessity that *inertia* has its origin in a kind of *interaction* of the bodies, completely in accordance of your thoughts on Newton's bucket experiment.

Note in particular the end of the first sentence and the beginning of the second: *If* the equivalence principle is confirmed, *then* so is the relativity of inertia.

Of course, this inference holds only if *both* the equivalence principle and the claim that gravitation is an interaction hold. There is a short publication in which Einstein replies to criticism from Ernst Reichenbächer from 1920,[83] the year of his intense exchange with Schlick, where Einstein is explicit in regard to the claim that gravitation is an interaction, with the clarification that the interaction is said to be mediated by $g_{\mu\nu}$. The outcome, incidentally, is a revised description of the 1916 thought experiment of the two rotating spheres:

> Mr. *Reichenbächer* misunderstood my considerations regarding two celestial bodies rotating with respect to one another. One of these bodies is rotating in the sense of Newtonian Mechanics, and thus flattened by centrifugal effects, the other is not. This is what the inhabitants would measure with rigid rods, tell each other about it, and then ask themselves about the real cause of the different behaviour of the celestial bodies. (This has nothing to do with Lorentz contraction.) *Newton* answered this question by declaring absolute space real, with respect to which one but not the other allegedly rotates. I myself am of the Machian opinion, which in the language of relativity theory can be put in the following way: All masses of the world together determine the $g_{\mu\nu}$-field, which is, judged from the first celestial body, a different one than judged from the second one; for the motion of the masses producing the $g_{\mu\nu}$-field differ significantly. Inertia is, in my opinion, a (mediated) interaction between the masses of the world in the same sense as those effects which in Newtonian theory are considered as gravitational effects.

A more concise statement of the interaction claim was made in 1921,[84] in the context of providing an argument supporting what we today call the Einstein universe, i.e., the universe in which Mach's Principle would still hold despite de Sitter's result:

> [I]t seems, therefore, natural to trace back the complete inertia of a body to an interaction between itself and the other bodies in the world, just as since Newton gravitation is traced back completely to an interaction between the bodies.

To summarise, it seems fair to say that Einstein did not need a variant of the action–reaction principle as a reason to adopt the relativity of inertia in 1913. His strong belief in the equivalence between gravity and inertia, together with his retention of the Newtonian tenet that gravity is an interaction between bodies, could be seen as reason enough.[85] Furthermore, it is the pairing of the equivalence principle and the principle of the relativity of inertia, together with the principle of relativity, that Einstein mentions repeatedly up until 1920 as the cornerstones of GR, whereas AR only really takes centre stage in 1920 in the correspondence with Schlick and in subsequent publications. For these reasons, we are inclined to believe that the 1920 correspondence brought out a watershed in Einstein's thinking, marking an unprecedented shift in Einstein's interpretation of the superiority of GR over preceding theories of space-time: Its superiority now rested on satisfaction of the action–reaction principle, rather than implementation of Mach's original analysis of inertia.

Final remarks

Einstein's frequent references to GR's vindication of the action–reaction principle in the years following his 1921 Princeton lectures have been noted in a number of studies.[86] A particularly telling quotation is from a letter Einstein wrote a year before his death to Georg Jaffe:

> You consider the transition to special relativity as the most essential thought of relativity, not the transition to general relativity. I consider the reverse to be correct. I see the most essential thing in the overcoming of the inertial system, a thing which acts upon all processes, but undergoes no reaction. The concept is in principle no better than that of the centre of the universe in Aristotelian physics.[87]

For Einstein, the glory of GR rested partly on its alleged superiority to preceding theories of space-time that involve absolute structure. His 1924 essay 'On the ether' contains a particularly clear denunciation of Newtonian mechanics in terms of its violation of AR.[88] But caution should be exercised when extrapolating backwards, as it were, in the history of physics. It does not automatically follow from the fact that GR satisfies AR and that NM and SR do not, as we mentioned in the first section above. To repeat, Einstein was content in his 1905 development of SR to explicitly borrow the inertial frames from NM, without any fretting about the correct metaphysics of action. Of course, if AR is to be respected in these theories, inertia must be taken as a brute fact, a position advocated, in different ways, by Schlick and others, as we have seen. Such a position is surely defensible in the context of these theories.

The two epigrammatic Einstein quotations cited at the beginning of this essay underscore how Einstein's thinking changed between 1905 and 1913, and again between 1913 and 1924. In the years 1912 and 1913, when Mach's influence on him may have been greatest, Einstein had convinced himself that the phenomenon of inertia required a causal explanation, while regarding as absurd the notion of immaterial space acting as such a cause. By 1924, he was stressing that the metric field in GR is as real and efficacious as the electromagnetic field, and in particular could indeed be seen as the origin of inertia. (But it is worth stressing here that Einstein did not view GR as furnishing a *geometric* explanation of gravitational phenomena; he continued to reject the notion of space, or space-time, as providing the cause of inertia.)[89]

Nowadays, acceptance of Einstein's 1924 claim should be seen to rest not simply on the nature of $g_{\mu\nu}$ and its geodesics, but rather on the so-called *geodesic theorem*, which demonstrates that the form of Einstein's field equations, along, it must be noted, with other plausible universal assumptions about matter fields, imply that the world lines of test particles are time-like geodesics as defined by the metric field.[90] Note that the theorem deals with an idealisation; it states that extended, but *truly* freely falling bodies only *approximately* move inertially.[91] In fact, it is a subject worthy of investigation as to whether the details of the theorem are strictly consistent with Einstein's insistence that a violation of AR holds in theories with absolute space-time structure.[92] But such an investigation must be pursued elsewhere. It is our hope that in the present chapter, some further light has been shed on the circumstances that led Einstein to bring to the fore the role of the action–reaction principle in his new theory of gravity.

Acknowledgments: R.B. thanks Partha Ghose and Peter Ronald deSouza, Director of the IIAS, for the invitation to speak at the 2012 conference 'The Nature of Reality: The Perennial Debate'. He thanks Julian Barbour for very the helpful discussions of the first draft of the chapter, in which a number of mistakes were noted and suggestions for improvements were made. He also acknowledges fruitful discussions with Paul Lodge and Gonzalo Rodriguez-Pereyra concerning Leibniz's theory of causation. D.L. thanks the Einstein Papers Project, in particular Diana Kormos Buchwald and Tilman Sauer, for training, without which his part in this work would not have been possible, and the Center for Philosophy of Science of the University of Pittsburgh for their hospitality during part of the work on the chapter. He also wishes to thank John Norton in particular, for his patience and for inspiring discussions on Einstein and the interpretation of Einstein's work. Finally, both authors thank John Norton, Tilman Sauer and particularly Oliver Pooley for their detailed comments on an earlier draft of the chapter, which led to a number of improvements.

Notes

1 Vol. 5, Doc. 495, *The Collected Papers of Albert Einstein* (CPAE for short from now on). All translations from Einstein texts are based on the corresponding translation volumes of the CPAE unless otherwise noted; in some cases, slight modifications of translation have been introduced.

2 Einstein, Albert, 'U¨ ber den A¨ ther', *Schweizerische Naturforschende Gesellschaft. Verhandlungen* 105 (1924). English translation by Simon Saunders, in Saunders, S. and Brown, H. (eds.), *Philosophy of Vacuum* (Oxford: Oxford University Press, 1991), pp. 13–20, 85–93, 88. Translation by DL.

3 For an insightful analysis of the Einstein-Tagore dialogues, see Home, D. and Robinson, A., 'Einstein and Tagore: Man, nature and mysticism', *Journal of Consciousness Studies* 2(2) (1995), pp. 167.

4 See Einstein, Albert, 'Autobiographical notes', in *Albert Einstein, Philosopher-Scientist*, Paul Arthur Schilpp (ed.) Vol. 7. *The Library of Living Philosophers* (La Salle, IL: Open Court Publishing, 1949), pp 1–94. For related remarks, see

Einstein, A., 'Quantenmechanik und Wirklichkeit', *Dialectica* 2(3–4) (1948), pp. 320–24, reprinted and translated as 'Quantum mechanics and reality', in *The Born-Einstein Letters: Correspondence Between Albert Einstein and Max and Hedwig Born from 1916 to 1955* (New York: Walker, 1971). Very helpful analyses of Einstein's philosophical position(s) can be found in

Howard, D., 'Realism and conventionalism in Einstein's philosophy of science: The Einstein-Schlick correspondence in philosophy of science history of science', *Philosophia Naturalis* 21(2–4) (1984), pp. 616–29,

Howard, D., 'Was Einstein really a realist?', *Perspectives on Science* 1(2) (1993), pp. 204–51,

Fine, A., *The Shaky Game: Einstein, Realism, and the Quantum Theory* (Chicago: University of Chicago Press, 1996),

Dongen, Jeroen van, *Einstein's Unification* (Cambridge: Cambridge University Press, 2010), chapter 2, and

Norton, J. D., 'Philosophy in Einstein's science', in *Alternatives to Materialist Philosophies of Science*, Philip MacEwen (ed.) (The Mellen Press, forthcoming).

5 A discussion of the 'principle of causality' is particularly prominent in the Einstein-Schlick correspondence, as reviewed in the section of this name below. However, note that at the time it was quite common not to draw a clear distinction between causality and determinism; in many cases, demanding that the 'principle of causality' holds is equivalent to demanding that every event has a determining cause or set of causes. For our discussion, the important point is that in the beginning of the 1920s, Einstein started to think of the metric field as (causally) determining things, rather than just being determined by the distribution of masses.

6 In a Lagrangian framework (which Einstein started to use extensively from 1918 onwards), this mutual affection can be represented by the metric and the matter fields (both dynamical) *coupling* to each other. The stress-energy tensor, however, turns out to be a relational property of the matter fields, which they possess in virtue of their relations to the metric field. For a recent discussion

of the relational significance of the stress-energy tensor, see Lehmkuhl, D., 'Mass-energy-momentum: Only there because of spacetime?', *The British Journal for the Philosophy of Science* 62(3) (2011), pp. 454–88; section 4.3 for different kinds of coupling.

7 For recent versions of this argument, see Brown, H. R. and Pooley, O., 'Minkowski space-time: A glorious non-entity', in *The Ontology of Spacetime*, Dennis Dieks (ed.) Vol. I. *Philosophy and Foundations of Physics* (Boston: Elsevier, 2006), and Brown, H. R., *Physical Relativity. Space-Time Structure from a Dynamical Perspective* (Oxford: Oxford University Press, 2007).

It should be noted that these analyses differ from Newton's, in the sense that there is no commitment to the reality of space as a fundamental entity.

8 Vol. 5, Doc. 495, CPAE.

9 Schlick to Einstein, 10 June 1920, Vol. 10, Doc. 51 CPAE.

10 A translation is found in Hall, A. R. and Hall, M. B., *Unpublished Scientific Papers of Isaac Newton: A Selection from the Portsmouth Collection in the University Library* (Cambridge: Cambridge University Press, 1962). The date of this important manuscript is still a matter of dispute. There is little doubt that it was written between 1666 and 1685; for details of the conflicting views on the likely date and a defence of the early claim, see Henry, J. 'Gravity and *De Gravitatione*: The development of Newton's ideas on action at a distance', *Studies in History and Philosophy of Science Part A* 42(1) (2011), pp. 11–27.

11 Ibid.

12 Leibniz made this point in a 1700 letter to Bernoulli; see Gerhardt, C. I., *Leibnizens mathematische Schriften. Leibnizens Gesammelte Werke* (Halle: Verlag von H. W. Schmidt, 1858), Vol. III.

13 See, for example, Ishiguro, Hidé, 'Pre-established harmony versus constant conjunction: A reconsideration of the distinction between rationalism and empiricism', *Proceedings of the British Academy* LXIII (1993), pp. 239–62 and Woolhouse, R. S., 'Pre-established harmony retuned: Ishiguro versus the tradition', *Studia Leibnitiana* (1985), pp. 204–19.

14 See, for example, Brown, G., 'Is there a pre-established harmony of aggregates in the Leibnizian dynamics, or do non-substantial bodies interact?', *Journal of the History of Philosophy* 30(1) (1992), pp. 53–75 for a careful discussion of the question of whether pre-established harmony applies in Leibniz's philosophy at the level of aggregate substances.

15 See Squires, E. J., 'Some comments on the de Broglie-Bohm picture by an admiring spectator', in *Waves and Particles in Light and Matter*, A. van der Merwe and A. Garuccio (eds.) (New York: Plenum, 1994), pp. 125–138.

16 See Anandan, J. and Brown, H. R., 'On the reality of space-time geometry and the wavefunction', *Foundations of Physics* 25(2) (1995), pp. 349–60.

17 Einstein, A., *Vier Vorlesungen über Relativitätstheorie gehalten im Mai 1921 an der Universität Princeton*. Reprinted as Vol. 7, Doc. 71 CPAE; and in various editions as 'The meaning of relativity' (Princeton, NJ: Princeton University Press and Braunschweig: F. Vieweg, 1922).

18 See Stein, H., 'Newtonian space-time', *Texas Quarterly* 10 (1967). Reprinted in Palter, Robert (ed.), *The Annus Mirabilis of Sir Isaac Newton 1666 to 1966* (Cambridge, MA: MIT Press, 1970), pp. 258–84, 174–200,

Barbour, J., *The Discovery of Dynamics: A Study from a Machian Point of View of the Discovery and the Structure of Dynamical Theories* (New York: Oxford University Press, 2001), Chapter 11,

Rynasiewicz, R., 'By their properties, causes and effects: Newton's scholium on time, space, place and motion—I. The text', *Studies in History and Philosophy of Science Part A* 26 (1995), pp. 133–53,

Rynasiewicz, R., 'By their properties, causes and effects: Newton's scholium on time, space, place and motion—II. The context', *Studies in History and Philosophy of Science Part A* 26 (1995), pp. 295–321, and

Pooley, O., *The Reality of Spacetime* (Oxford University Press, forthcoming), Chapter 2.

19 Newton, I., 'De Gravitatione', in *Unpublished Scientific Papers of Isaac Newton*, A. Hall and M. Hall (eds.) (Cambridge: Cambridge University Press, 1962), section 7. The inspiration for Newton's view that space is neither substance nor accident may have been the Renaissance thinker Francesco Patrizi da Cherso, or indeed Walter Charleton. See, in particular, Pooley, note 18, section 2.1, for a discussion of possible influences on Newton in this regard. A translation of and commentary on Patrizi's 'De Spacio Physico' can be found in Francesco Patrizi. 'On physical space', *Journal of the History of Ideas* 4 (1943). Translated by Benjamin Brickman, pp. 224–25.

20 *de Grav* (see note 7 above), section 5.

21 See particularly Barbour, J. 'The part played by Mach's Principle in the genesis of relativistic cosmology', in *Modern Cosmology in Retrospect*, S. Bergia, A. Messina, B. Bertotti and R. Balbinot (eds.) (Cambridge: Cambridge University Press, 1990), pp. 47–66,

Norton, J. D. 'Mach's principle before Einstein', in *Mach's Principle: From Newton's Bucket to Quantum Gravity*, Julian Barbour and Herbert Pfister (eds.) Vol. 6. *Einstein Studies* (Boston: Birkhäuser, 1995), p. 9–57

Hoefer, C. (a), 'Einstein's formulations of Mach's principle', *ibid*, pp. 67–90,

Hoefer, C. (b), 'Einstein's struggle for a Machian gravitation theory', *Studies in History and Philosophy of Science* 25(3) (1994), pp. 287–335,

Renn, Jürgen, 'The third way to general relativity: Einstein and Mach in context', in *The Genesis of General Relativity*, Jürgen Renn and Matthias Schemmel (eds.) Vol. 3, *Boston Studies in the Philosophy of Science* (Dordrecht: Springer, 2007).

22 Norton, 'Mach's Principle before Einstein', note 21.

23 Barbour, J., 'General relativity as a perfectly Machian theory', in *Mach's Principle: From Newton's Bucket to Quantum Gravity*, Julian Barbour and Herbert Pfister (eds.) (Boston: Birkhäuser, 1995), pp. 214–36, section 2.

Note that this paper also defends a version of Machianism that, unlike Mach's own version, is consistent with (certain solutions of) the field equations of GR.

24 Both of these figures were interested in non-Euclidean geometry and eschewed Riemann's separation of physics from geometry. (For further discussion of their views, see

Kragh, H. (a), 'Geometry and astronomy: Pre-Einstein speculations of non-euclidean space', *arXiv* 1205(4909) (2012),

Kragh, H. (b), 'Zöllner's universe', *Physics in Perspective* 14(4) (2012), pp. 392–420, where it is pointed out that in 1872, Zöllner proposed a cosmological model describing a finite universe in closed space. A connection between Zöllner's suggestion of a cosmological underpinning of the force law in electrodynamics and 'later versions of Mach's principle' is suggested in Kragh (a), 'Geometry and astronomy', p. 26. It is unclear to us, however, just to what extent these views are fully consistent with the action–reaction principle.

25 Mach, E., *Space and Geometry in the Light of Physiological, Psychological and Physical Inquiry* (Chicago and London: The Open Court Publishing Company, 1906), pp. 136 and 141. See in this connection Kragh (b), 'Zöllner's universe', p. 35.

26 Ibid.

27 Norton, 'Mach's principle before Einstein', note 21, p. 28.

28 Mach, E., *Die Geschichte und die Wurzel des Satzes von der Erhaltung der Arbeit*, Translation as 'History and root of the principle of the conservation of energy', P. Jourdain (ed.) (Chicago: Open Court, 1911). Prag: Calve'sche Buchhandlung, 1872, p. 61; Mach's emphasis.
29 Barbour, 'The part played by Mach's Principle in the genesis of relativistic cosmology', note 21, p. 48.
30 *Op. cit.*, pp. 49–50. For further analysis, see in particular Hoefer (a), note 21.
31 Einstein, A., 'Gibt es eine Gravitationswirkung, die der elektrodynamischen In- duktionswirkung analog ist?', *Vierteljahrschrift für gerichtliche Medizin und öffentliches Sanitätswesen* 44 (1912). Reprinted as Vol. 4, Doc. 7 CPAE, pp. 37–40.
32 Einstein finally abandoned this theory primarily because he convinced himself that the theory was in conflict with Newton's third law; see
Einstein, A., 'Zur Theorie des statischen Gravitationsfeldes', *Annalen der Physik* 38 (1912). Reprinted as Vol. 4, Doc. 4 CPAE, pp. 443–58, 452–58; and see
Norton, J. D., 'Eliminative induction as a method of discovery: How Einstein discovered general relativity', in *The Creation of Ideas in Physics: Studies for a Methodology of Theory Construction*, J. Leplin (ed.) *The Western Ontario Series in Philosophy of Science* (Dordrecht: Kluwer Academic Publishers, 1995), pp. 29–70, section 5.1, for details.
33 Einstein, A. and Grossmann, M., *Entwurf einer verallgemeinerten Relativitätstheorie und einer Theorie der Gravitation, Zeitschrift für Mathematik und Physik*, 62 (1913) 225–244. Reprinted as Vol. 4, Doc. 13 CPAE.
34 Einstein, 'Gibt es eine Gravitationswirkung, die der elektrodynamischen In- duktionswirkung analog ist?', note 31, p. 38.
35 Ibid.
36 Einstein, A., 'Zum gegenwärtigen Stande des Gravitationsproblems', *Physikalische Zeitschrift* 14 (1913). Reprinted as Vol. 4, Doc. 17 CPAE, pp. 1249–66, 1260.
37 Note that Einstein did not clearly distinguish between the relativity of inertia [the predecessor of 'Mach's principle' as defined only in
Einstein, A., 'Prinzipielles zur allgemeinen Relativitätstheorie', *Annalen der Physik* 55 (1918), Reprinted as Vol. 7, Doc. 4 CPAE, pp. 241–44] and the relativity of motion, as he himself admits of not having done up until 1918. For details on different versions of these principles and the development in Einstein's thought, see the sources summarised in note 21.
38 See Vol. 5, Docs. 448 and 495, CPAE.
39 In the letter, this becomes clear when Einstein directs Mach's attention to a particular page in the *Entwurf* paper (p. 6), where he explicitly claims that the inertial mass of a body is a function of the gravitational potential in the static limit of the *Entwurf* theory, which he then relates to Mach's 'bold idea' of a relativity of inertia.
40 Einstein, A., 'Die formale Grundlage der allgemeinen Relativitätstheorie', *Sitzungsberichte der Königlich Preussischen Akademie der Wissenschaften* 1914 (part 2), 1030–1085. Reprinted as Vol. 6, Document 9 CPAE, pp. 799–801.
41 Norton, J. D., 'General covariance and the foundations of general relativity: Eight decades of dispute', *Reports on Progress in Physics* 56(7) (1993), pp. 791, 810.
42 See ibid., section 3.1.
43 When Einstein failed to find convincing generally covariant field equations, he convinced himself of their impossibility with the now famous 'hole argument', only to return to the requirement of general covariance in launching the 'point-coincidence argument' in 1915. See
Stachel, J., 'Einstein's search for general covariance', in *Einstein and the History of General Relativity*, Don Howard and John Stachel (eds.) Vol. 1. *Einstein Studies* (Boston:Birkhäuser, 1989), pp. 63–100,

Earman, J. and Norton, J., 'What price substantivalism? The hole story', *British Journal for the Philosophy of Science* 38 (1987), pp. 515–25,

Norton, 'General covariance and the foundations of general relativity: Eight decades of dispute', note 41, and Pooley, *The Reality of Spacetime*, note 18 for details, including references to criticism of Einstein's equating the 'general principle of relativity' with general covariance.

44 Barbour, 'The part played by Mach's Principle in the genesis of relativistic cosmology', note 21, p. 53.

45 Renn, Jürgen and Sauer, Tilman, 'Pathways out of classical physics. Einstein's double strategy in his search for the gravitational field equations', in *The Genesis of General Relativity*, Jürgen Renn (ed.) Vol. 1, *Einstein's Zurich Notebook: Introduction and Source Introduction and Source* (Dordrecht: Springer, 2006), p. 301.

46 Einstein, A., 'Die Grundlage der allgemeinen Relativitätstheorie', *Annalen der Physik* 49(7) (1916). Reprinted as Vol. 6, Doc. 30 CPAE, pp. 769–822.

47 Ibid.

48 Howard, 'Was Einstein really a realist?', note 4, p. 211.

49 Einstein, 'Die Grundlage der allgemeinen Relativitätstheorie', note 46.

50 See Brown, note 7, Appendix A. Einstein's views on the meaning of general covariance would undergo considerable changes. Already in Einstein, Albert, 'Hamiltonsches Prinzip und allgemeine Relativitätstheorie', *Sitzungsberichte der Königlich Preussischen Akademie der Wissenschaften* 1916 (part 2), 1111–1116, reprinted as Vol. 6, Doc. 41 CPAE, he would emphasise its role in clarifying the conservation principle in GR, in providing what was a special case of Noether's second 1918 theorem; see ibid and especially

Janssen, M. and Renn, Jürgen, 'Untying the knot: How Einstein found his way back to field equations discarded in the Zurich notebook', in *The Genesis of General Relativity*, Jürgen Renn (ed.) Vol. 2 (Dordrecht: Springer, 2006), section 3.

See also Norton, 'General covariance and the foundations of general relativity: eight decades of dispute', note 41, especially section 5, for the development of the role of general covariance in Einstein's thought.

51 Barbour, 'The part played by Mach's Principle in the genesis of relativistic cosmology', note 21, p. 48.

52 Einstein, A., 'Ernst Mach', *Physikalische Zeitschrift* 17 (1916). Reprinted as Vol. 6, Doc. 29 CPAE, pp. 101–4.

53 See Einstein and Grossmann, note 33, p. 6.

54 Einstein, *Vier Vorlesungen über Relativitätstheorie gehalten im Mai 1921 an der Universität Princeton*, note 17, pp. 64–6.

55 See Kretschmann, E., 'Über den physikalischen Sinn der Relativitätspostulate. A. Einsteins neue und seine ürsprüngliche Relativitätstheorie', *Annalen der Physik* 358(16) (1918), pp. 575–614.

56 Einstein, 'Prinzipielles zur allgemeinen Relativitätstheorie', note 37.

57 This was noted in note 6 above; the dire implications for the 1918 Mach's Principle were spelt out recently in Lehmkuhl, 'Mass-energy-momentum', note 6, p. 482.

58 See Barbour, J. and Pfister, H. (eds.), *Mach's Principle: From Newton's Bucket to Quantum Gravity*. Vol. 6, *Einstein Studies* (Boston: Birkhäuser, 1995), p. 93.

59 Einstein, A., 'Kosmologische Betrachtungen zur allgemeinen Relativitätstheorie', *Sitzungsberichte der Königlich Preussischen Akademie der Wissenschaften* 1917 (part 1), 142–152. Reprinted as Vol. 6, Doc. 43, CPAE.

60 Einstein initially tried to attack de Sitter's solution on various grounds, arguing that it was i) not static (unacceptable to Einstein at the time) and ii) that it involved an intrinsic singularity and thus matter. In the end, Einstein had to

admit that it was an entirely viable mathematical solution to the modified field equations (involving no intrinsic singularities). He went on to rule out the solution as unphysical because it was not globally static; a move he repeated when Friedman, A., 'Über die Krümmung des Raumes', *Zeitschrift für Physik A Hadrons und Nuclei* 10(1) (1922), pp. 377–86 published his non-static solution and regretted painfully after E. Hubble ('A relation between distance and radial velocity among extra-galactic nebulae', *Proceedings of the National Academy of Sciences* 15(3) (1929), pp. 168–73) published his results on the redshift of galaxies. See

Frenkel, V., 'Einstein and Friedmann', in Yuri Balashov and Vladimir Vizgin (eds.), *Einstein Studies in Russia*, Vol. 10. *Einstein Studies* (Boston: Birkhäuser, 2002) pp. 1–15; for Einstein's attack on Friedman's work and subsequent developments,

Einstein, A. and De Sitter, W., 'On the relation between the expansion and the mean density of the universe', *Proceedings of the National Academy of Sciences of the United States of America* 18(3) (1932), p. 213

for Einstein's reaction to Hubble's results, and the editorial note 'The Einstein-de Sitter-Weyl-Klein debate' of Vol. 7, CPAE for more on the exchange between Einstein and de Sitter, as well as Janssen, Michel, ' "No Success like Failure . . .": Einstein's Quest for General Relativity, 1907–1920', in *The Cambridge Companion to Einstein*, Michel Janssen and Christoph Lehner (eds.) (Cambridge: Cambridge University Press, 2014), pp. 167–227.

61 Einstein, note 54, p. 36.

62 Einstein, *Äther und Relativitätstheorie*, note 26, pp. 11–3.

63 The fact that Einstein uses the word 'ether' extensively in ibid. seems more due to the circumstances in which the paper was delivered than to the content Einstein wanted to bring the term across; Einstein rarely used the term in later writings when referring to $g_{\mu\nu}$ or $F_{\mu\nu}$, but sometimes said that whether one says 'space-time' or 'ether' does not really matter in the end. The fact that Einstein used the term excessively in ibid., however, must be related to it being his inaugural lecture in Leiden, which Lorentz and Ehrenfest had engineered and which Einstein used to express his reverence for Lorentz.

64 See the introduction to Schlick, Moritz, *Moritz Schlick Gesamtausgabe, Abteilung 1: Veröffentlichte Schriften, Band 1*, Fynn Ole Engler and Matthias Neuber (eds.) (Vienn: Springer, 2006) for a detailed biography of Schlick, alongside corresponding references.

65 Einstein to Schlick, 14 December 1915, Vol. 8, Doc. 165, CPAE.

66 Schlick, M., *Raum und Zeit in der gegenwärtigen Physik* (Berlin: Springer, 1917), p. 178. Translation by DL.

67 Vol. 8, Doc. 314 CPAE.

68 Schlick, M., 'Naturphilosophische Betrachtungen u¨ber das Kausalprinzip', *Die Naturwissenschaften* 8(24) (1920), pp. 461–74.

69 See Howard, 'Realism and conventionalism in Einstein's philosophy of science', note 4,

Hentschel, K., 'Die Korrespondenz Einstein-Schlick: zum Verhältnis der Physik zur Philosophie', *Annals of Science* 43(5) (1986), pp. 475–88 and

Hentschel, K., *Interpretationen und Fehlinterpretationen der speziellen und der allgemeinen Relativitätstheorie durch Zeitgenossen Albert Einsteins* (Basel: Birkhäuser, 1990), especially section 4.7.3, for more on the Einstein-Schlick debate.

70 See Call No. 21-575 EA for the corresponding letter, and Vol. 5, p. 576 CPAE for a summary of the letter.

71 Einstein to Schlick, 7 June 1920, Vol. 10, Doc. 47 CPAE.

72 Schlick to Einstein, 10 June 1920, Vol. 10, Doc. 51 CPAE.
73 Vol. 10, Doc. 67 CPAE.
74 Vol. 10, Doc. 116 CPAE.
75 Vol. 10, Doc. 47 CPAE.
76 Schlick to Einstein, 10 June 1920 (Vol. 10, Doc. 51 CAPE).
77 See Holton, G., 'Mach, Einstein, and the search for reality', *Daedalus* 97(2) (1968), pp. 636–73 and van Dongen, note 4, pp. 37–40.
78 Norton, 'General covariance and the foundations of general relativity: eight decades of dispute', note 41, p. 810.
79 Einstein, 'Prinzipielles zur allgemeinen Relativitätstheorie', note 36, pp. 1260–61.
80 It is true that Einstein rejected his own 1912 scalar field theory (mentioned in note 32 above) when he discovered that it failed to satisfy Newton's third law of motion concerning action–reaction. But this is a case of the existence of *both* action and reaction, which happen not to be equal and opposite, thus giving rise to an unacceptable force-free accelerative phenomenon. As we stressed in section 2, AR is not to be conflated with Newton's third law, which is a much stronger constraint on the way bodies act on each other.
81 Hoefer (b), note 21, especially in section 1.1, argues that Einstein's conception of the equivalence principle between 1907 and 1912 already contains an 'implicit Machianism'. We hope to clarify in this section how the equivalence principle and the relativity of inertia were likely related to each other in Einstein's mind.
82 Einstein uses this term, for example, in a letter to Lorentz from 18 February, 1912, (Vol. 5, Doc. 360, CPAE) when discussing the alleged equivalence between the two. Note, however, that Einstein puts 'acceleration fields' in quotes himself.
83 Einstein, A., 'Antwort auf Ernst Reichenbächer, "Inwiefern läßt sich die moderne gravitations-theorie ohne die Relativität begründen?"', *Die Naturwissenschaften* 8 (1920). Reprinted as Vol. 7, Doc. 49, CPAE, pp. 1010–11.
84 Einstein, A., *Geometrie und Erfahrung* (Berlin: Springer, 1921). Reprinted as Vol. 7, Doc. 52, CPAE; see also Vol. 7, Doc. 31 CPAE for a similar statement from December 1919/January 1920.
85 Compare Norton, J., 'What was Einstein's principle of equivalence?', in *Einstein Studies in History and Philosophy of Science, 16* (1985), pp. 203–246; reprinted in D. Howard and J. Stachel (eds.), *Einstein and the History of General Relativity: Einstein Studies* Vol. I, Boston: Birkhauser, 1989, pp. 5–47: '[I]t was natural for expect that the extended theory, which dealt with general gravitational effects, would explain the observed disposition of inertial frames of reference in terms of the matter distribution of the universe. For the structure that determined this disposition would behave in many aspects like a traditional gravitational field and therefore be strongly influenced by any motion of its sources, the masses of the universe'.
86 See in particular Norton, note 41 and Hoefer (1995), note 21.
87 See Stachel, J., 'Einstein and the quantum: fifty years of struggle', in Robert G. Colodny (ed.), *From Quarks to Quasars: Philosophical Problems of Modern Physics* (Pittsburgh: University of Pittsburgh Press, 1986), pp. 349–81, 377.
88 Einstein, 'U¨ ber den A¨ ther', note 2, p. 88.
89 For details of Einstein's arguments against seeing GR as a 'geometrization of gravity', see

Lehmkuhl, D., 'Why Einstein did not believe that general relativity geometrizes gravity', *Studies in the History and Philosophy of Modern Physics* 46 (2014), pp. 316–26; for related arguments, see

Anderson, James I., 'Does general relativity require a metric', *arXiv preprint gr-qc/9912051* (1999) and

Brown, H. R., 'The behaviour of rods and clocks in general relativity, and the meaning of the metric field', *arXiv Preprint: 0911.4440* (2009). Forthcoming in the *Einstein Studies Series*.

90 See Malament, D. B. 'A remark about the "Geodesic Principle" in general relativity', in Melanie Frappier, Derek Brown and Robert DiSalle (eds.), *Analysis and Interpretation in the Exact Sciences: Essays in Honor of William Demopoulos* (Dordrecht: Springer, 2012), pp. 245–52.

91 See Brown, *Physical Relativity*, note 7, p. 141, and particularly

Tamir, M., 'Proving the principle: Taking geodesic dynamics too seriously in Einstein's theory', *Studies in History and Philosophy of Modern Physics* 43 (2012), pp. 137–54.

92 See the brief comments in Brown, *Physical Relativity*, note 7, pp. 141, 142.

2 The hole argument and the nature of space-time

A critical review from a constructivist perspective

Tian Yu Cao

Introduction

In the great conversation on the Nature of Reality held between Rabindranath Tagore and Albert Einstein in mid-July of 1930,[1] Einstein's clearly stated view that 'Truth must be conceived as a Truth that is valid independent of humanity' is in sharp contrast with Tagore's equally clearly stated view that 'the Truth of the Universe is human Truth'. In this chapter, the nature or the truth of space and time, which is an important part or aspect of reality, is explored from a perspective based on the constructive version of structural realism.[2] More specifically, Einstein's deliberations on the so-called hole argument in the years leading to his 1915 formulation of the general theory of relativity and their philosophical implications, especially those concerning the nature of space and time, will be critically as well as historically and scientifically examined. That is, they will be examined in the context of the dominant views on space and time held by Newton, Leibniz, Kant, Engels, Mach and those suggested by recent developments in the studies of quantum gravity. The result of our exploration suggests that the case of the truth of space-time appears to vindicate a synthesis of the views espoused by Tagore and Einstein.

The hole argument

John Stachel's authoritative interpretation of Einstein's hole argument and its sequels leading to the genesis of the general theory of relativity is influential and well known.[3] Thus, a brief summary is enough for the later discussions.

The hole argument, according to Stachel, says that general covariance has deprived manifold points' individuality if causality is not to be compromised. This implies that points would have no spatial-temporal meaning until they are enmeshed into a relational structure as placeholders whose identity is determined by their places and roles in the structure. The relational structure referred to here is constituted by metric tensor fields, which are solutions to the field equation. More on the relationship between this relational structure and the physically substantial field will be given soon. A further implication is that the states of the metric field are not labeled by manifold points (or supported by manifold points, as the substantivalist philosopher of space-time John Earman[4] frequently claims), but can be characterised, for example, by a field's internal structural features, the first and second fundamental forms defining scalar and external curvatures. The argument's implication for the nature of space-time being dynamic and relational is contained in Einstein's 1952 claim that space-time has no separate existence, but only as the structural quality of the field.[5] This claim has made physicists and philosophers busy in digesting and explaining. Lee Smolin and Carlo Rovelli have stressed its relationist implications.[6] Stachel's reading is more complicated. He has succinctly summarised the lesson learnt from the hole argument: no kinematics without dynamics. This claim, I think, should be read properly without improper inflation or deflation. Geometry certainly sits at the core. But what about dimension, boundary conditions and other fundamental parameters? Should we aspire to derive all of them from a dynamic theory? The answers to these questions are closely related with one's understanding of the nature of space-time, which is the subject of the next section.

Traditional views on the nature of space-time

Traditionally, the nature of space-time is understood in two ways.[7] The substantivalist school presumes an ontologically independent existence of space-time, which serves as the stage for the physical entities, events and processes. The major arguments for this position, according to Earman, are, first, the necessity of a reference framework for accommodating absolute motions, and second, the necessity for any field theory to have an ontological support. These two arguments will be addressed soon. According to the anti-substantivalist school, space-time is a derivative structure, which has to be explained by or derived from ontologically primary physical entities, relations and processes.

Newton's absolute space is a substantival entity. It functions as a reference frame and is supposed to be the precondition for absolute motion to be conceivable. It was conceived by Mach as a monster because it acted on physical objects without being acted upon by them.[8] It can be argued, as Earman did, as an ontological foundation for field theories. The argument is valid for classical field theories as well as relativistic quantum field theories based on the Minkowskian space-time. But the validity collapses once we take the general theory of relativity into consideration, as the hole argument has indicated.

For Leibniz, space and time were not ontologically independent entities, but just the totality of spatial and temporal relations among objects. Here, the relations were not conceived as ontologically self-subsistent non-supervenient relations without being anchored on relata, as Paul Teller[9] and ontic structural realists Steven French and his followers[10] have frequently claimed, but presumed as the existence of material objects. Leibniz's relationist version of anti-substantivalism was based on two principles: the principle of sufficient reason and that of the identity of the indiscernible, or PII. There is no sufficient reason, however, for having to have a sufficient reason for everything. If you push it hard enough, it will reveal itself as just a disguised version of theology. As to PII, anyone who gets acquainted with quantum physics would discard it immediately.

Mach's anti-substantivalist view of space was different. While Newton insisted that the ultimate reference framework, with which absolute motions can be defined, should be conceived as an absolute space, Mach assumed that the required reference framework was in fact provided by the configuration of physical degrees of freedom in the universe. Thus, we have obtained a functional absolute space without having committed to an ontologically independent absolute space.[11]

Mach's conception of the reference framework opened the door for Einstein to introduce an agent, the gravitational field, which, due to its universal coupling to all physical degrees of freedom in the whole universe, is able to provide the required reference framework in specifying spatial and temporal relations without itself being the space-time. The implications of introducing the gravitational field, which is dynamic in nature, into the understanding of space-time are rich and deep, and have been exploited by Einstein first and many others afterwards, John Stachel included.

A crucial difference between Einstein's view of the spatial and temporal relations and the relationist view held by Leibniz and all the way down to Adolf Grunbaum[12] is that these relations, in Einstein's case, are not externally specified, but rather are intrinsically constituted by the metric tensor field. That is why Einstein claimed that space-time is only a structural quality of the field. Starting from this, we can develop a structural, constitutive and constructive view of space-time, to which I now turn.

A constructivist approach

If the metric tensor field plays such a crucial role in Einstein's view of space-time, a problem naturally arises is how to characterise these fields. Do they exist in space-time? Where to define the equation for the metric tensor field? Surely, we need a manifold. Even in more fundamental theories, the same question remains. In many theories of quantum gravity, such as those of spin foams, causal sets, causal dynamical triangulations and in geometrogenesis, no space-time is presumed; it only emerges in the classic limit from the quantum theory. In these theories, all the spatial-temporal kinematic structures can be deduced from local dynamic processes. But the dynamic processes still have to be conceptualised in terms of a manifold. In this respect, a leader in loop quantum gravity, Abhay Ashtekar, once argued against Julian Barbour, a fundamentalist in the relationist school, in the late

1980s, that 'one cannot even write the constraint equations if one does not have a manifold as such'.[13]

So we have to start form a manifold, without which no field equation would be formularisable and no sensible conception of space-time that is informed by the best available scientific theories would be possible. But is it legitimate to turn this epistemic necessity of using a manifold into an ontological argument in support of the so-called manifold substantivalism, as Earman has argued? What is exactly the ontological status of a manifold?

A reasonable answer to this question can be based on the assumption that the totality and structures of the points in a manifold can be viewed as *isomorphic* to those of the totality of physical events in the whole universe. It may even have the status of a pre-space-time, similar to Kant's space and time as *a priori* forms of our sensibility that can be used to structure our experience and make sense of our experiences. One possible argument in support of this claim is that the dimensionality of a four-dimensional manifold is deeply anchored in our perceptual structure.

Even though a minimally structured manifold (with only a global topological structure and dimensionality) is the starting point for further construction of the reality of space-time, this acknowledgment of its epistemic necessity cannot be exploited in support of manifold substantivalism for two reasons. First, as we have already pointed out, the manifold points have no direct spatio-temporal meaning, and their space-time meaning is constituted by the chromo-geometrical structures (metric and the closely related Riemann tensor), which endow the points with individuality either through imposing non-reflexive metrical relationships upon manifold points as argued by Simon Saunders in 2003,[14] or by characterising the points with four or more invariants of the Riemann tensor, as indicated by Bergmann in 1957,[15] Komar in 1958[16] and Stachel in 1993. For this reason, the points of a manifold enjoy no ontological priority over the chromo-geometrical structures. Second, as Stachel indicated a long time ago, even the global structures of a manifold are not fixed and globally defined, but a result of the maximal compatible extension of local solutions to a dynamic theory; thus, a loophole to a manifold substantivalism is closed.

The above-mentioned two reasons can be rephrased in terms of the relationship between a field and a manifold. First, if in classical field theories or in special relativistic field theories, such as conventional quantum field theories, fields are anchored on the points of a manifold, this is not the case in the general theory of relativity. Here, as Einstein stressed in 1952, it is not that fields are situated in a space-time manifold, but rather these fields are spatially extended and temporary endured and endow the manifold points with spatial and temporal meanings. Thus, a manifold is deprived of its ontological *a priori* status, not to speak of its space-time connotation, by the field. A manifold is constituted by certain fields into a space-time with certain structural features. Second, if the global structure of a manifold as space-time is the result of the global extension of local solutions to a field equation, a manifold can only be legitimately viewed as the mathematical expression of a field's spatial and temporal features.

A further construction of the reality of space-time is facilitated by a clarification of the relationship between kinematic and dynamic structures. Here we have to notice that kinematic structures, mainly the chromo-geometrical structures, are not substantial but purely relational. Thus, substantial entities have to be found to ontologically support them, so that the dynamics of these structures can be materialistically rather than Platonically understood. The gravitational field, as a substantial physical entity represented by the connection field, is the required dynamical entity, whose relational aspects are mathematically represented by the metric tensor, which represents the chromo-geometrical structure. The two are inseparably connected to each other through the geodesic equation as the compatibility condition. Thus, although the spatio-temporal relations are specified by the metric, the metric itself is ontologically supported by the inertio-gravitational field (the connection). Once the relationship between the metric and the connection is thus clarified, the mystery about the dynamicity of the metric (as a purely relational structure), or the dynamic nature of space-time in general, as is convincingly put in display through the general theory of relativity, is dispelled: It is only an *epiphenomenon* of the dynamical behaviour of the connection, as a substantial physical structure or entity.

Both the metric and connection are taken to be holistic structures that enjoy ontological priority over their components. For the metric, manifold points are only placeholders for the spatio-temporal relations it stipulates; for the connection, it is both a substantial entity with universal couplings with all kinds of physical entities and a holistic structure stipulating the possible behaviours of test bodies if they are put somewhere in the field and interact with parts of the field (the local values of the field) there. Clearly, this structural, constitutive and constructive view of space-time, a new version of the anti-substantivalist view of space-time, distinguishes itself from the externalist relationism with its intrinsic constitution of spatial-temporal relations by the metric and with its clarification of the ontological relationship between the substantial gravitational field and the metric field as the mathematical expression of its relational structure.

A dialectical ascendancy from the abstract to the concrete

The major steps in our construction of the reality of space-time can be summarised as follows. First, we started from a Kantian form of sensibility, an unstructured bare manifold as a pre-space-time whose points could be used to represent physical point-coincident events, although these points themselves have no spatial and temporal meanings. In a sense, these points have quiddity of space-time in a generalised Kantian sense. For example, the dimensionality of the manifold is closely related with human perceptive structure. But these points do not have haecceity of space-time before being enmeshed with a metric tensor field; that is, they cannot differentiate themselves from each other and thus are unable to support spatial-temporal relationships, which can only be constituted by manifold points with haecceity. This lack of haecceity of manifold points is only a mathematical expression of the fact that the individuality of these points as the anchor of the spatial-temporal events is constituted by a gravitational field. With the introduction of the metric tensor field, not only the Leibnizian relations, but also the functions of the Newtonian absolute space-time can be effectively and convincingly derived or reconstructed.

Second, with the clarification of the ontological relationship between the metric field and the gravitational field, we get a clearer sense that the reality of space-time is ultimately constituted by the physically substantial gravitational field. The introduction of this dynamical agent into the constitution of space-time has an additional benefit for the constructivist view as against the substantivalist view because according to the latter, the absolute space-time can only be an empty reference framework without any source of dynamicity.

Once a series of substantial and dynamical fields are introduced for the constitution of various real and concrete incarnations of space-time, the door for the exploration of the richness of space-time is opened. One may think about Elie Cartan and torsion,[17] but I mainly have quantum gravity in mind. When quantum gravity is taken into consideration, the advantage of the constitutive approach becomes supreme. The emergence of classical space-time from the quantum regime through chains of phase transitions is heterogenerous in nature. This means that qualitatively novel, different space-times can result due to different dynamic processes that intrinsically determine those phase transitions. So the concrete and rich reality of space-time can be investigated through the investigations of concrete physical processes connecting the quantum regime and the classical regime.[18]

Third, this constitutive view has brought us from Kant's form of sensibility as the starting point of construction of the reality of space-time to Friedrich Engels's conception of space and time as forms of existence of matter.[19] Engels's notion of matter here can be properly understood as physical degrees of freedom. Engels's form seems to have affinity to Kant's form. But there is a leap from the subjective to the objective. The dialectics of form and content and the unity of form and content would allow us to explore the roles of physical entities and processes in constituting the reality of space-time, the exploration of which I just surveyed moments ago.

Now the various stages in the constitution are coincident with the steps in our construction of the reality of space-time, from a very abstract Kantian form of pre-space-time to a more and more concrete reality of space-time. These steps can be viewed as the steps in the ascendance from the abstract to the concrete in Hegel's dialectical logic, which was used by Karl Marx in his construction of social reality. Since what are involved in the constructive ascendancy described above are not the logical categories but physical entities, events and processes, this dialectical constructive approach to understanding the nature of space-time should be regarded as an application of Marx's dialectics rather than Hegel's.

Then what is the concrete reality of space-time we have constructed so far? In my view, space-time as a means of structurally describing the behaviour of physical degrees of freedom under the influence of gravitation is essentially an ontological expression of gravity's effects. Turning attention from space-time to the dynamical issue of gravity was the major contribution of Einstein. It is, to my knowledge, also the consensus among physicists and most philosophers of space-time.

The Tagore-Einstein synthesis on the nature of space-time

The above examination suggests that the reality of space-time is the result of stepwise human construction: from the very abstract Kantian form of pre-space-time to a more and more concrete reality of space-time, which seems to be in line with Tagore's wisdom. On the other hand, although these constructive steps can be viewed as the steps in the ascendance from the abstract to the concrete in Hegel's dialectical logic, what are involved in the constructive ascendancy are not the Hegelian logical categories, but *physical entities, events and processes*. For this reason, the dialectical constructive approach to understanding the nature of space-time should be regarded as an application of Marx's materialistic dialectics rather than Hegel's idealist one; that is, the truth of space-time is a human construction, which, however, is based on objective reality. This conclusion seems in line with Einstein's 'religious belief in super-human objectivity'.

Acknowledgment: An early draft was presented at the colloquium 'An Intellectual Life Across Disciplines' in honour of John Stachel's 85th birthday on 12 September, 2013 in Berlin, Germany. I am grateful to the Max Planck Institute for the History of Science for organising the event and giving me the opportunity to receive feedback from Professor John Stachel and other participating and non-participating audience members.

Notes

1 Tagore and Einstein (1930): a conversation between Rabindranath Tagore and Professor Albert Einstein on the afternoon of 14th July, 1930, at the professor's residence in Kaputh (From *The Religion of Man* [Appendix II], supplemented from the typescripts in the archive at Rabindra Bhavana and *The Kenyon Review*, New Series 23(2), *Cultures of Creativity: The Centennial Celebration of the Nobel Prizes* [Spring, 2001], pp. 20–5, published by Kenyon College). See also Appendix 1 of this volume.

2 Cao, T. Y., 'Structural realism and the interpretation of quantum field theory', *Synthese* 136(1) (2003), pp. 3–24; Cao, T. Y., 'Structural realism and quantum gravity', in *Structural Foundation of Quantum Gravity, Dean Rickles*, Steven French and Juha Saatsi (eds.) (Oxford: Oxford University Press, 2006), pp. 42–55; Cao, T. Y., *From Current Algebra to Quantum Chromodynamics—A*

Case for Structural Realism (Cambridge: Cambridge University Press, 2010), Chapter 9.

3 Stachel, J., 'Einstein's search for general covariance, 1912–1915', a paper presented to *the 9th International Conference on General Relativity and Gravitation*, Jena, 1980; later printed in *Einstein and the History of General Relativity*, D. Howard and J. Stachel (eds.) (Boston: Birkhäuser 1989), pp. 63–100; Stachel, J., 'The meaning of general covariance', in *Philosophical Problems of the Internal and External Worlds*, J. Earman, A. Janis, G. Massey and N. Rescher (eds.) (Pittsburgh: University of Pittsburgh Press 1993), pp. 129–60.

4 Earman, J., *World-Enough and Space-Time* (Cambridge, MA: MIT Press 1989).

5 Einstein, A., 'Relativity and the problem of space', appendix 5 in the 15th edition of *Relativity: The Special and the General Theory* (London: Methuen 1954), pp. 135–57.

6 See Smolin, L., *Three Roads to Quantum Gravity* (London and New York: Weidenfeld and Nicolson and Basic Books 2001) and Rovelli, C., *Quantum Gravity* (Cambridge: Cambridge University Press 2004).

7 The traditional views are briefly yet precisely summarised in Earman (1989).

8 Mach, E., *Die Mechanik in ihrer Entwicklung. Historisch-Kritisch dargestellt* (Leipzig: Brockhaus 1883); see also Cao, T. Y., *Conceptual Development of 20th Century Field Theories* (Cambridge: Cambridge University Press 1997), section 4.3.

9 Teller, P., 'Relational holism and quantum mechanics', *British Journal for the Philosophy of Science* 37 (1986), pp. 71–81.

10 French, S. and Ladyman, J. (a), 'Remodelling structural realism: Quantum physics and the metaphysics of structure', *Synthese* 136(1) (2003), pp. 31–56; French, S. and Ladyman, J. (b), 'Between platonism and phenomenalism: Reply to Cao', *Synthese* 136 (2003), pp. 73–8.

11 See Cao, *Conceptual Development of 20th Century Field Theories*, section 4.3 on Mach.

12 Grunbaum, A., 'Absolute and relational theories of pace and space-time', in *Foundations of Space-Time Theories*, J. Earman, C. Glymore and J. Stachel (eds.) (Minneapolis: University of Minnesota Press 1977), pp. 303–73; see also Cao, *Conceptual Development of 20th Century Field Theories*, p. 121.

13 More on this and relevant references can be found in Cao, 'Structural realism and quantum gravity'.

14 Saunders, S., 'Indiscernibles, general covariance, and other symmetries: The case for non-reductive relationalism', in *Revisiting the Foundations of Relativistic Physics*, A. Ashtekar, J. Renn, L. Divarci, P. Schroeter, R. Cohen, D. Howard, S. Sarkar and A. Shimony (eds.) (Dordrecht: Kluwer, 2003), pp. 151–73.

15 Bergmann, P. G., 'Topics in the theory of general relativity', in *Brandeis University Summer Institute of Theoretical Physics*, E. N. Parker, J. S. Goldstein, A. A. Maradudin and V. Ambegaokar (eds.) (New York: W. A. Bejamin, 1957), pp. 1–44.

16 Komar, A., 'Construction of a complete set of independent observables in the general theory of relativity', *Physical Review* 111 (1958), pp. 1182–87.

17 Cartan, E., 'Sur une géneralisation de la notion de courbure de Riemann et les éspaces à torsion', *Comptes Rendus De L Academie Des Sciences Paris* 174 (1922), pp. 593–95.

18 Cao, T. Y., 'Conceptual issues in quantum gravity', *Invited Lecture at the 13th International Congress of Logic, Methodology and Philosophy of Science*, 9–15 August 2007, Beijing China, unpublished.

19 Engels, F., *Dialectics of Nature* 1883 (New York: International Publishers, April 1968).

3 Quantum information and reality, especially the reality of the past and future

Charles H. Bennett

Information is quantum: how physics has helped explain the nature of information and the origin of randomness

Like other parts of mathematics, the theory of information originated as an abstraction from everyday experience: calculation is a manipulation of pebbles and a digit is a finger or a toe. The information revolution is based on these abstractions, crystallised in the mid 1900s by Turing, Shannon, von Neumann and others. But these notions are now known to be too narrow. Quantum theory, developed by physicists in the early 1900s and spectacularly successful in its own field, also provides a more complete and natural arena for developing concepts of communication and computation.

Conventionally, information carriers have been viewed as what a physicist would call *classical* systems:

(a) Their states in principle are reliably distinguishable, and can be observed without disturbing the system.
(b) To specify the joint state of two or more systems, it is sufficient to specify the state of each one separately.

But for quantum systems like atoms or photons:

(a) Attempting to observe a particle's state in general disturbs it, while obtaining only partial information about the state (uncertainty principle).
(b) Two particles can exist in an *entangled* state, in which they behave in ways that cannot be explained by supposing that each particle has some state of its own.

For most of the 20th century, quantum effects in information processing were regarded mainly as a nuisance, because the uncertainty principle makes quantum devices behave less reliably than the classical ideal. Now it is known that quantum effects also have positive consequences, making possible new kinds of information processing, such as quantum cryptography, and dramatically speeding up some classically hard computations. These positive effects are chiefly due to entanglement. Moreover, entanglement helps explain why quantum effects are so inconspicuous and remained undiscovered until the 20th century.

Despite the differences, there are important similarities between classical and quantum information. All (classical) information is reducible to bits 0 and 1. All processing of information, to reveal hidden truths, can be done by simple logic gates (AND, NOT) acting on one bit and two bits. Bits and gates are *fungible* in the sense of being independent of their particular physical embodiment, making possible Moore's law.

Quantum information is reducible to *qubits*, i.e. two-state quantum systems such as a photon's polarisation or a spin-1/2 atom. Quantum information processing is reducible to one- and two-qubit gate operations. Qubits and quantum gates are fungible among different quantum systems.

Differences between classical and quantum information

Ordinary classical information, such as one finds in a book, can be copied at will and is not disturbed by reading it. Quantum information is more like the information in a dream.

Trying to describe your dream changes your memory of it, so eventually you forget the dream and remember only what you have said about it. You cannot prove to someone else what you dreamed. You can lie about your dream and not get caught. But unlike dreams, quantum information obeys well-known laws.

The superposition principle

The central principle of quantum mechanics is the *superposition principle*. Between any two reliably distinguishable states of a physical system (for example, vertically and horizontally polarised single photons), there are intermediate states (for example, diagonal photons) that are not reliably distinguishable from either original state.

The possible physical states of a system correspond to directions in a space. The dimensionality of this space is equal to the system's maximum number of reliably distinguishable states.

Any direction is a possible state, but two states are reliably distinguishable if only if their directions are perpendicular. For a more mathematical treatment, see the *Appendix*.

Using polarised photons to carry information

Photons behave reliably if measured along an axis parallel or perpendicular to their original polarisation. Used in this way, each photon can carry one reliable bit of information. But measuring the photons along any other axis causes them to *behave randomly*, forgetting their original polarisation direction.

A rectilinear (i.e. vertical vs. horizontal) measurement distinguishes vertical and horizontal photons reliably, but randomises diagonal photons.

A diagonal measurement distinguishes diagonal photons reliably but randomises rectilinear photons.

No measurement can distinguish all four kinds. This is not a limitation of particular measuring apparatuses, but a fundamental consequence of the uncertainty principle. This fundamental limitation gives rise to the possibility of quantum money and quantum cryptography. *Quantum money*[1] cannot be copied by a counterfeiter, but can be checked by the bank, which knows the secret sequence of polarised photons it should contain. *Quantum cryptography* uses polarised photons to generate shared secret information between parties who share no secret initially[2].

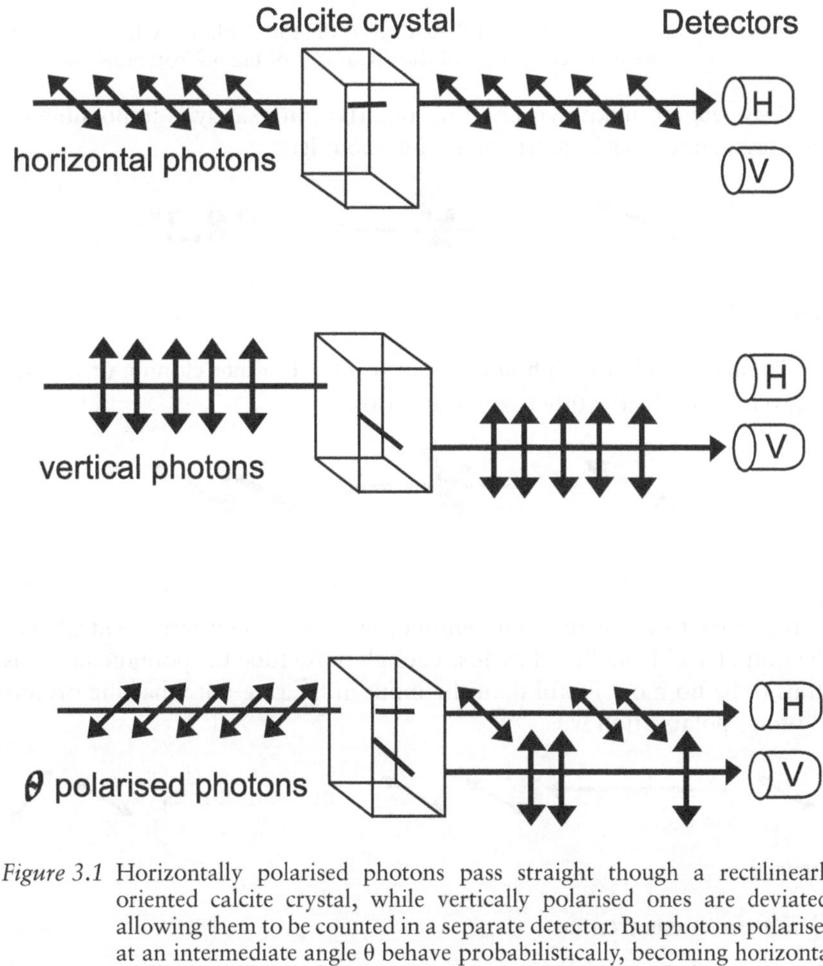

Figure 3.1 Horizontally polarised photons pass straight though a rectilinearly oriented calcite crystal, while vertically polarised ones are deviated, allowing them to be counted in a separate detector. But photons polarised at an intermediate angle θ behave probabilistically, becoming horizontal with probability $\cos^2\theta$ and vertical with probability $\sin^2\theta$.

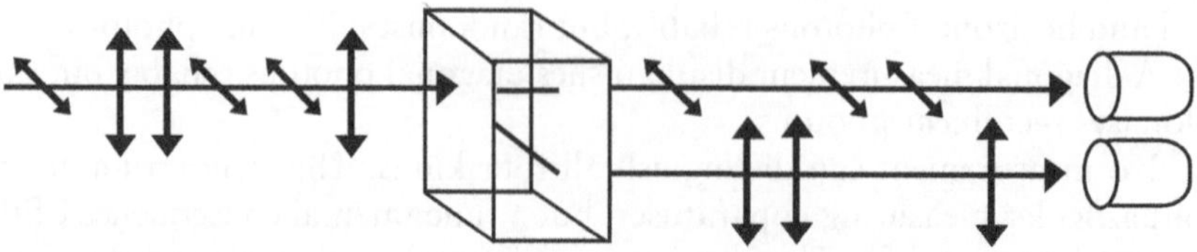

Figure 3.2 Distinguishing horizontal and vertical photons by a rectilinear measurement.

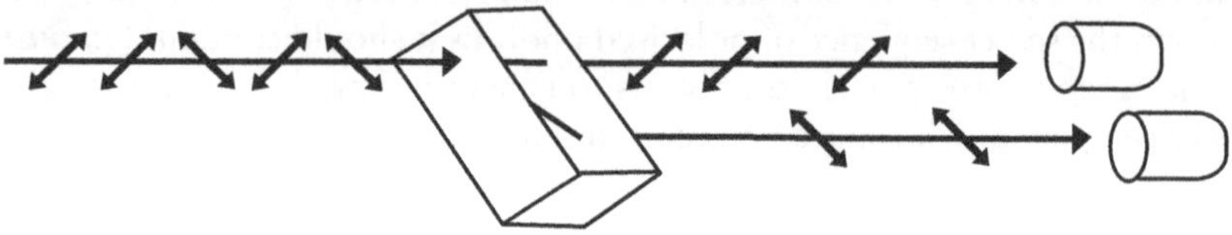

Figure 3.3 Distinguishing 45 and 135 degree diagonal photons by a diagonal measurement, consisting of the apparatus of Fig 3.2 rotated 45 degrees.

Measuring an unknown photon's polarisation exactly is impossible (no measurement can yield more than 1 bit about it).

Figure 3.4

Cloning an unknown photon is impossible. (If either cloning or measuring were possible, the other would be also).

Figure 3.5

If you *try* to clone an unknown photon by sending it into an ideal laser, the output will be polluted by just enough noise (due to spontaneous emission) to be no more useful than the input in figuring out what the original photon's polarisation was.

Figure 3.6

Quantum measurement

Quantum physicists often use the words *observation* and *measurement* as near synonyms, describing situations in which a system under study interacts with another system, animate or inanimate, so that the latter learns or records something about the former. When a classical system is observed, e.g. looking at the moon through a telescope or a bacterium through a microscope, to a good approximation the observer gains information about the system under study without disturbing it or by disturbing it only a little. Quantum physics thoroughly undermines this common-sense separation, showing that it is only approximate and apparent, rather like the pre-Einsteinian independence of space and time, but in a way that is even more counterintuitive and hard to put into everyday language. Let us first describe the phenomenology of quantum measurement, and then discuss how to explain it and overcome its unfamiliarity.

The best pedagogic metaphor for quantum measurement I know is that suggested by Prof. William Wootters of Williams College, USA. He compares it to the student-teacher interaction in an old-fashioned traditional school, where students were not permitted to ask questions but were only expected to listen and answer the teacher's questions. The student is the photon, and the teacher is the measuring apparatus consisting of the calcite crystal and the two photon detectors.

——*Prof. Wootters's Pedagogic Metaphor for Quantum Measurement*——

Like a pupil confronting a strict teacher, the quantum system being measured is forced to choose from among a set of distinguishable states characteristic of the measuring apparatus.

Teacher:	Is your polarisation vertical or horizontal?
Pupil:	Uh, I am polarised at about a 55-degree angle. . . .
Teacher:	**I believe I asked you a question.** Are you vertical or horizontal?
Pupil:	Horizontal, sir.
Teacher:	Have you ever had any other polarisation?
Pupil:	No, sir. I was always horizontal.

Einstein never made his peace with this picture, always believing that some deeper, smoother theory must underlie the seeming abruptness and randomness of this process, in which the measured system forgets its earlier state and jumps to one of several new ones, which one being absolutely

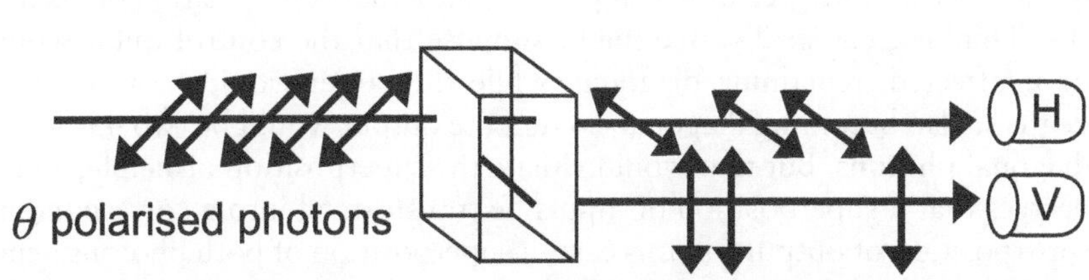

Figure 3.7 Like a strict teacher, a rectilinear measuring apparatus forces photons to choose to be horizontal or vertical, forgetting their original polarization.

Figure 3.8 Quantum CNOT gate: If the upper 'control' qubit is horizontal, the lower 'target' qubit is unaffected. If the control qubit is vertical, the target is rotated 90 degrees.

unpredictable beforehand. Later, I will show how the dichotomy between measurer and measured system can be harmonised, and the seeming abruptness and randomness to a considerable extent is explained using the notion of entanglement, which itself is an unavoidable consequence of the superposition principle.

How entanglement arises from interactions between quantum systems

Recall that in pointing out the similarities between quantum and classical information processing, we noted out that any quantum data processing can be done by 1- and 2-qubit gates acting on qubits. Let us consider one of the simplest kinds of interaction between qubits, the quantum version of the familiar classical XOR operation on two bits. In a classical XOR gate, the first or 'control' bit is unaffected, while the second or 'target' bit gets flipped if and only if the control bit was 1. Using the Dirac half-angle bracket notation for quantum states, where |0> and |1> represent two standard, distinguishable states of a qubit, or in the case we are considering, the horizontal and vertical states of a polarised photon, the quantum XOR gate (often called a 'controlled-NOT' . . . or CNOT gate) rotates the target 90 degrees if the control is vertical, but leaves the target alone if the control is horizontal.

Fig. 3.8 shows how this gate would behave in two situations, with a vertical or horizontal control qubit applied to a horizontal target qubit.

We now ask how the CNOT would behave on a superposition of these two situations, with the target qubit being horizontal (|0>) but the control qubit being a 45-degree diagonal photon, the equal superposition of |0> and |1>. Thinking classically, one might suppose that the control qubit would be unaffected, remaining diagonal, while the target would be rotated 45 degrees, also becoming diagonal, so that the output would be two 45-degree diagonal photons. But that would violate the superposition principle, which decrees that a superposition of inputs be transformed into a corresponding superposition of outputs, in this case, a superposition of both photons being vertical and both being horizontal.

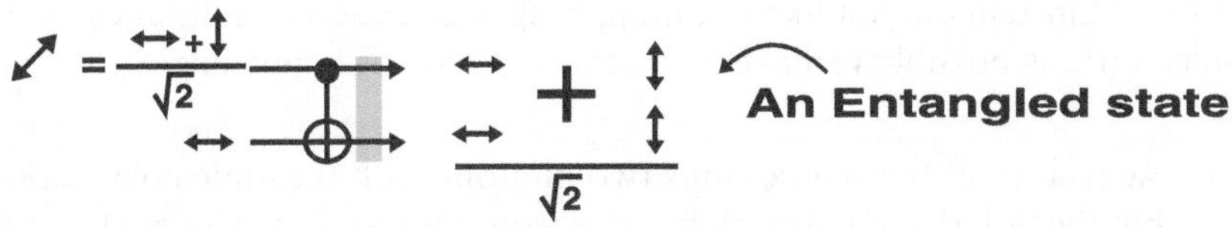

Figure 3.9

What is this state? Since it is a state of two photons, each of which has two reliably distinguishable states, it lives in a space of four dimensions, whose four axes can be identified with the four distinguishable states |00>, |01>, |10> and |11>, where the first symbol in each bracket refers to the upper (control) photon and the second refers to the lower (target) photon. By the superposition principle, it must be the particular direction illustrated above, or in symbols (|00> + |11)/√2, which lies midway between both photons being horizontal and both being vertical. This is a different direction in four-dimensional space from both photons being diagonal and indeed different from *any* state in which each photon has a definite polarisation of its own. It is an *entangled* state, meaning a definite state of a compound system that cannot be described by ascribing definite states to each part. By elementary algebra, the non-entangled state of both photons having 45-degree diagonal polarisation can be shown to be (|00> + |01> + |10> + 11>)/2, which is a different direction in four-dimensional space.

Though different from any non-entangled state, the both-horizontal plus both-vertical entangled state is the same as an equal superposition of both photons being 45-degree diagonal and both 135-degree diagonal:

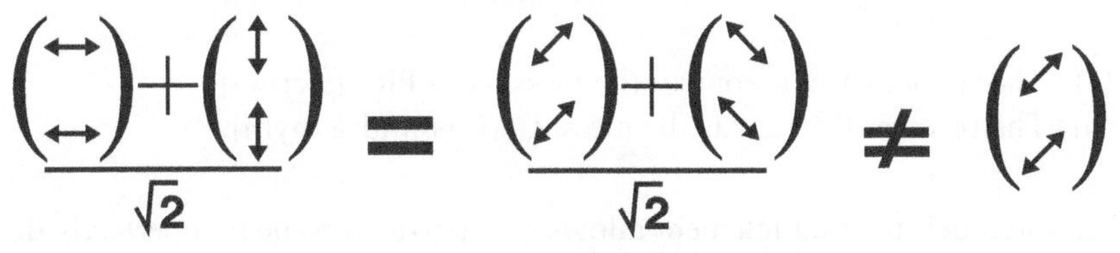

Figure 3.10

Thus, the two photons may be said to be in a state of sameness of polarisation, even though neither photon has a polarisation of its own.

Phenomenology of entanglement: behaviour of entangled photons when measured

If two photons are created in an entangled state, measuring either one along any axis gives a random result and simultaneously causes the other photon to acquire the same polarisation. This strange behaviour is called the EPR effect because it was first noted in a paper by Einstein, Podolsky and Rosen (EPR).[3] Einstein did not like it, calling it 'spooky action at a distance'. Consider various possible verbal explanations of what is happening:

(a) At each shot, the source emits two photons with the same polarisation, but the polarisation direction varies from shot to shot. This explanation fails because sometimes the source would emit two diagonal photons, and if these were both measured on the H/V axis, sometimes one would behave as V and the other as H. In fact, they always behave the same, both V or both H.
(b) Instantaneous action at a distance: This is also not possible because it violates special relativity, and besides, how does the first particle know where to send the message to?
(c) Quantum mechanics: Using the superposition principle and elementary algebra in four-dimensional spaces gives the right answer. But suppose you want to explain entanglement to a lay audience uncomfortable with algebra and four dimensions. Then I suggest a fourth way of describing entanglement in words, namely,
(d) A random, uncontrollable message backward in time.

This sounds like fuzzy thinking of the worst kind, because the ability to send messages backward in time would enable one to alter the past in such a way as to prevent one's own existence, the so-called grandfather paradox. But a 'message' backward in time is safe from paradox under two conditions, either of which frustrates your ability to kill your grandfather or advise your broker what stocks to buy or sell yesterday:

(a) The sender cannot control the message (EPR effect) OR
(b) The receiver disregards the message (Cassandra myth).

As noted before, entanglement allows two particles to be in a perfectly definite joint state, even though each by itself is completely random, like two hippies who know who they are, like, in perfect harmony, even though neither has an opinion on anything. Hippies believed that with enough LSD, everyone could be in perfect harmony with everyone else. Now we have a quantitative theory of entanglement and know that it is *monogamous*: the more entangled two systems are with each other, the less entangled they can be with anything else.

There is a pedagogic analog of entanglement, too: Twin pupils Remus and Romulus, who are each completely ignorant of all subjects, answering every question randomly, but always giving the same answer, even when questioned separately.

Teacher A: Remus, what colour is growing grass?
Remus: Pink, sir.
Teacher B (in another classroom): Romulus, what colour is growing grass?
Romulus: Pink, ma'am.

If such twins existed, they would be remarkable, but would have no capacity for communication. Likewise, entangled particles by themselves confer no ability to communicate.

Expressing classical data processing in quantum terms

Since our aim is to embed the classical theory of information processing in a larger quantum theory, we need to define a classical bit. A classical bit is just a qubit with one of the Boolean values 0 or 1. Similarly, a classical wire

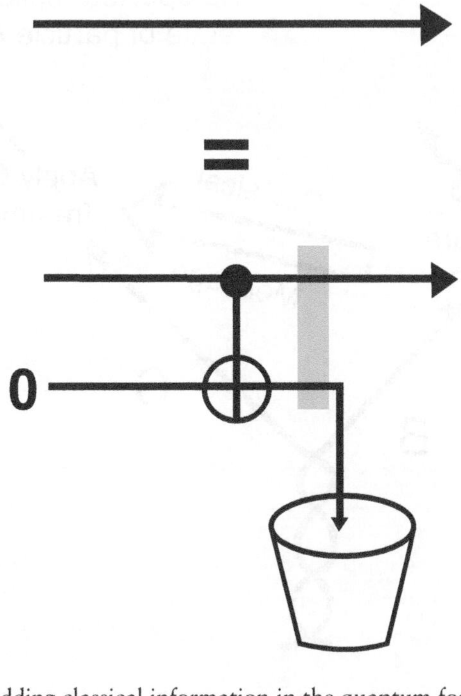

Figure 3.11 Embedding classical information in the quantum formalism: a classical wire is a quantum wire eavesdropped on by its environment (symbolized by the lower wire going into a wastebasket). If the input is 0 or 1, the environment receives a copy of it. But if it is a superposition, the output behaves randomly because it has become entangled with the environment.

(heavy arrow in the figure below) is a quantum channel that conducts 0 and 1 faithfully but randomises superpositions of 0 and 1. This happens because the data passing through the wire interacts with its environment, causing the environment to acquire a copy of it, if it was 0 or 1, and otherwise become entangled with it (indicated by shading in the figure). Thus, a classical channel is a quantum channel with an eavesdropper. A classical computer is a quantum computer handicapped by having eavesdroppers on all its wires.

Quantum teleportation

Given a particle in an unknown state, it would seem that the disruptive nature of quantum measurement would prevent complete information about the state from being read out of that particle and transferred to another particle that has never been anywhere near the first particle. Any attempt at measuring it might disturb it, still without extracting enough information

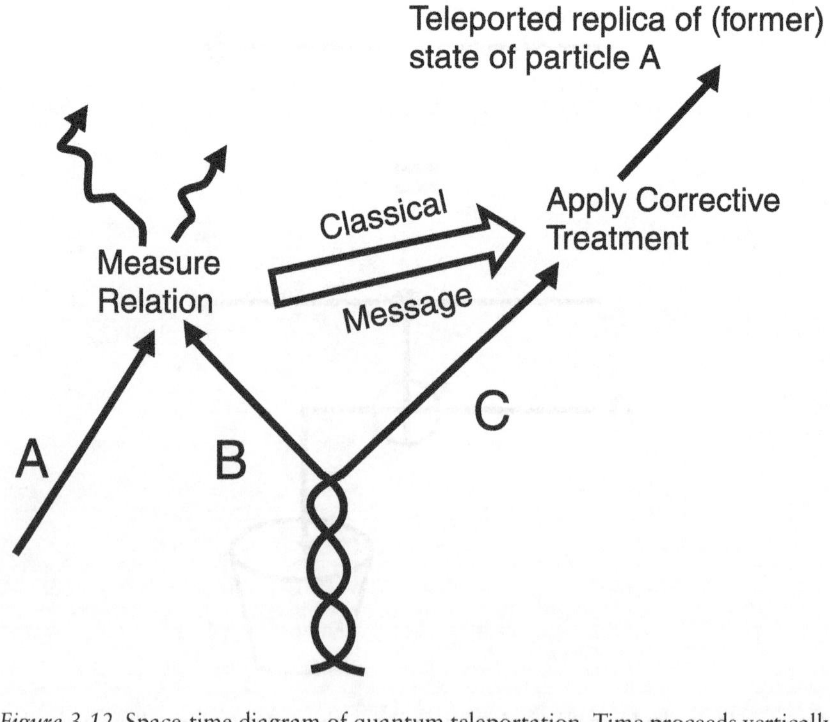

Figure 3.12 Space-time diagram of quantum teleportation. Time proceeds vertically upward. First particles B and C are prepared in an entangled state. Then a joint measurement is performed on particles A and B, forcing them into a random one of several entangled states. Which entangled state they have chosen is then communicated classically to the receiving location (wide arrow) where it is used to maneuver particle C into the same state particle A was in before it was measured.

to make a perfect replica. But quantum entanglement permits us to make an end run around this reasoning.

In quantum teleportation, the original unknown particle *A* in the sending location is not measured directly. Rather, a joint measurement is performed on it and another particle *B*, which has previously been prepared in an entangled state with particle *C*, which is sent to the receiving location. The measurement result is completely random and contains no information about the state of particle *A*, only about the relation between *A* and *B*, which of course had no relation initially but have been forced to choose one by the measurement. After the measurement result is transmitted classically to the receiving location, it is used to apply a corrective treatment to particle *C*, maneuvering *C* into exactly the same state *A* had been in before it was measured.

A human analogy for quantum teleportation

Suppose Alice has witnessed a complicated crime with possible terrorist implications where she lives, say, in Boston. The FBI in Washington knows that her memory of the crime is in a fragile form and does not want to ask her questions that might spoil it. They especially do not want to leave the investigation to the Boston police, who will ask her stupid questions and confuse her, so they invite her to Washington to be interviewed by a panel of experts who will ask her just the right questions in just the right order. Unfortunately, Alice dislikes travel and refuses to go. Fearing she will become uncooperative if they subpoena her, the FBI agrees to send one of their agents to Boston. But there is still a problem. The FBI experts all have strong opinions about the case and do not trust each other to conduct the interview alone. Finally, Remus volunteers, 'I know nothing about this case, so I am less likely to influence her than any of you. Besides, I like to travel. Just ask my brother'. Romulus concurs. So Remus goes to Boston to meet Alice. The meeting is a sort of speed date, with the parties instructed not to talk about anything substantive, just to concentrate on their relationship. The date goes badly, with Alice emerging a few minutes later saying, 'I can't stand him, and for some reason, this has all been so stressful that now I don't remember anything about the crime'. The Boston police thank Alice and tell her she can go home. Then they phone Washington and tell the FBI that Alice and Remus do not get along. The FBI experts go to Romulus and say, 'Well, it seems that Alice and your brother don't get on. So any question we would have asked Alice, we can ask you. We know that whenever you say yes, she would have said no'. They proceed with their careful questioning, reversing every one of Romulus's answers to get what Alice would have answered.

The ambiguity of the past and future

Reasoning from classical mechanics, Laplace thought the future and past were fully determined by the present, but attributed the perceived ambiguity of the future to our imperfect knowledge of the present, and/or our lack of sufficient computing power to calculate the future. An omniscient God would know past, present and future in full detail, although which moment we choose to call 'present' would seem rather arbitrary from the divine perspective. Quantumly, the future is less determined than Laplace imagined. Even an omniscient God would not be able to predict whether a particular radioactive atom will decay within its half-life. In our macroscopic world, we remember the past much better than we can predict the future. One can now scan all the books in Google Books to see how the frequencies of various phrases have varied over time. The phrase '1970' is mentioned rarely before that year, often immediately after, then with declining frequency.

Unlike the future, past macroscopic events are generally regarded as definite and unambiguous. Of course, some *microscopic* 'events' in the past (e.g., which path an unobserved photon followed through an interferometer) are regarded as being ambiguous, not because of ignorance, but because they are ill defined in principle. If either path through the interferometer is blocked, for example by the experimenter's finger, the photon leaves both exits equally often. Even when the photon does not strike the finger, its behaviour is altered by the finger's presence. But with both paths left open, the photon always leaves by the same exit, indicating that while passing unobserved through the apparatus, it followed a *superposition* of both paths. For it to have done so, no copy of the information of which path it followed can remain anywhere in the universe. One can say, after the experiment is over, that even God does not remember which path it followed. Or, as Daniel Greenberger put it, he would not believe in any god who claimed to know which path it followed.

As we shall presently argue, entanglement allows the evanescence of unobserved events in the past to be harmonised with the unpredictability of events in the future where, as in Wootters's pedagogic metaphor, a system is forced to jump randomly to one of a limited number of possible future states allowed by the measuring apparatus, forgetting its previous state.

These views can be harmonised by a property called the 'monogamy of entanglement', which, like entanglement itself, is a mathematical consequence of the superposition principle.

The monogamy of entanglement

Using personal intimacy as a metaphor for entanglement, if *A* and *B* are perfectly entangled with each other, they cannot be even classically correlated with anyone else. If *B* tries to share his entanglement with a third party, or lets it get eavesdropped on by the environment, his entanglement with *A* becomes degraded into mere classical correlation. If *B* then becomes perfectly entangled with *C*, he loses the classical correlation he once had with *A*. Thus, entanglements among particles follow the saying about entanglements among people: '*Two is a couple. Three is a crowd*'.

Most systems in nature, other than tiny ones like photons, interact so strongly with their environment that they soon become massively entangled with it. This destroys any previous entanglement that may have existed between internal parts of the system, degrading it into mere correlated randomness.

Entanglement and the origin of quantum randomness

Consider a diagonal photon about to be subjected to a measurement of its vertical versus horizontal polarisation. As noted, the result is a random outcome, with the photon forgetting its previous polarisation. But if no one observes the outcome of the measurement, the random 'behaviour' can be undone. This can be achieved by coherently recombining the separated beams before they enter the detectors by passing them through a polarisation rotator (for example, a piece of 'candy glass' used in the motion picture industry) and then a calcite crystal of equal thickness to the one that separated them in the first place.

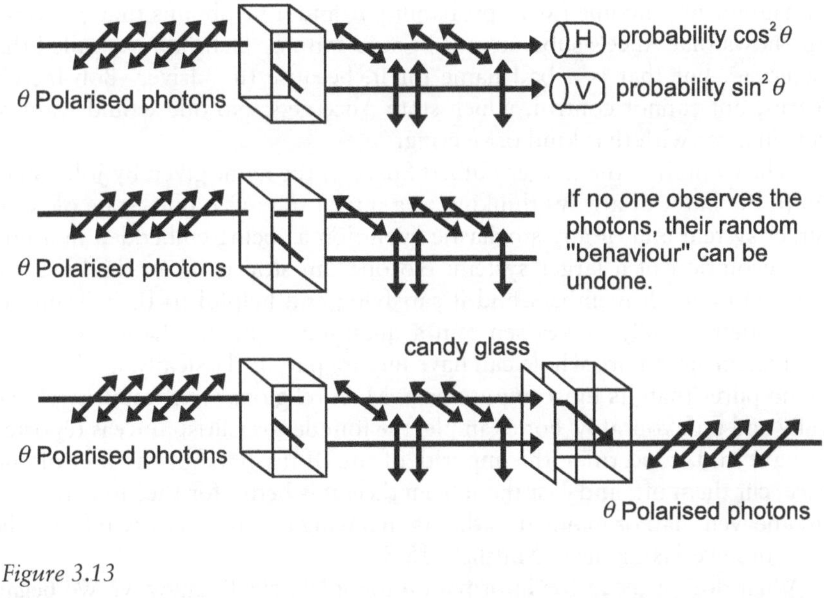

Figure 3.13

Speaking in terms of the metaphor of the strict teacher, it is not the teacher's question, but the *public embarrassment* of the pupil in front of the whole class, that makes him forget his original polarisation.

Meaning of the density matrix and the 'Church of the Larger Hilbert Space'

The quantum states we have spoken of so far, represented mathematically as directions in a suitable space, are called pure states because they represent situations of minimal ignorance, in which nothing more can be known about the system. Another notion, the mixed state, is used to represent situations of greater ignorance, in particular a probability distribution over an ensemble of possible pure states (analogous to classical ignorance), or the state of one part A of a compound system AB which itself is in a pure entangled state but of which the B part is inaccessible (a purely quantum kind of ignorance). Any mixed state is described mathematically by a unit-trace positive-semidefinite Hermitian density matrix ρ which establishes an elegant relation between the two kinds of ignorance.

The density matrix ρ represents all and only that information that can be obtained by sampling the ensemble or observing the A part of the compound system. Ensembles with the same ρ are indistinguishable. Pure states $\Psi(A,B)$ with the same $\rho(A)$ are indistinguishable by observing the A part. If Alice and Bob share a system in pure state $\Psi(A,B)$, then for any ensemble $\{\psi_i, p_i\}$ compatible with $\rho(A)$, there is a measurement that Bob can do on his subsystem alone, without Alice's knowledge or consent, that *remotely generates* the ensemble in her lab, in the sense that for each i, Bob gets measurement outcome i with probability p_i and if he obtains that outcome, he knows that Alice's subsystem is in pure state ψ_i. (Schrödinger called this 'steering', but that is a bad name for it, because the 'driver' Bob merely learns, but cannot control, which state Alice gets. No one would want to ride in a car with that kind of steering.)[4]

'The Church of the Larger Hilbert Space' is the name given by John Smolin to the habit of always thinking of a mixed state as a pure state of some larger system and of any stochastic evolution as being embedded in a unitary evolution of a larger system. No one can stop us from thinking this way, and church members find it satisfying and helpful to their intuition. This doctrine only makes sense in a quantum context, where because of entanglement, a pure whole can have impure parts. Classically, a whole can be no purer than its most impure part. Most religions view impurity classically and unfavourably. For example, the founder of Christianity is reported to have said, concerning the impurity of sin, 'If thy hand or thy foot offend thee, cut them off, and cast them from thee: it is better for thee to enter into life [heaven] halt or maimed, rather than having two hands or two feet to be cast into everlasting fire' (Matthew 18:8).

What does it mean for information to be 'classical', anyway? We began exploring this earlier, and continue in the next section.

Origins of classical phenomenology: Quantum Darwinism

The world we inhabit is far from a thermodynamic equilibrium. Not only are different macroscopic objects at different temperatures, but our world is immersed in a radiation field, a bath of photons of various wavelengths, very different from the uniform featureless blackbody radiation field that would prevail at thermal equilibrium. The near-equilibrium 3 Kelvin blackbody radiation left over from the Big Bang is swamped by far more numerous and energetic 5,000-Kelvin optical photons from the sun, which record the positions of objects on the surface of the earth in minute detail, making sight the most informative of our senses, and then, in enormous numbers, escape into space. Any object, even a micron-sized dust mote, illuminated by this radiation field gets its position redundantly recorded in escaping photons that will never again interact with the earth or with each other.

A qubit version of this process can be seen in the figure below, where the top qubit, symbolising a dust mote initially in a pure state ψ= α|0> + β|1>, interacts via CNOT gates with a sequence of zero qubits, representing parts of the environment that, like photons from the sun, interact with the system in a commuting fashion, but not with each other.

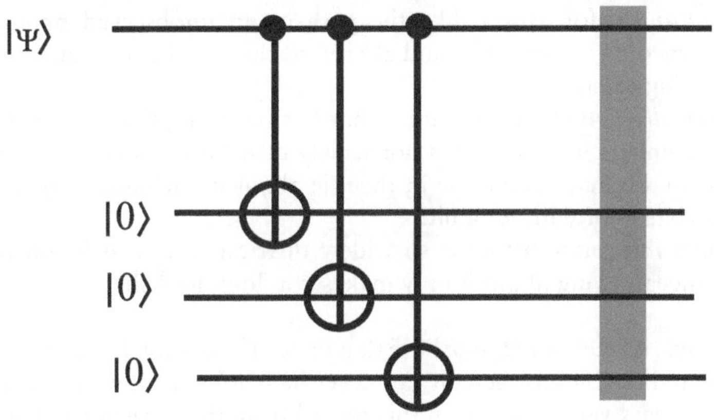

Figure 3.14

The result is a Schrödinger's-cat-like state α|**0***0000*..> + β|**1***1111*. . .>, in which the bolded first digit represents the system and the remaining italicised digits the subenvironments. In this state, there is a preferred measurement axis (|0> vs. |1>) along which the system is correlated with each subenvironment, while along complementary axes, it is correlated only with the whole environment. According to this paradigm, *information becomes classical by becoming public, that is, by being replicated redundantly throughout the environment, a process that (by the monogamy of entanglement) obfuscates complementary information by delocalising it among the subenvironments.* This approach to decoherence theory is called 'Quantum Darwinism',[5] though a better name might be 'Quantum Spam', because the numerous 'offspring' are all copies of a single original, rather than evolving with variation and selection through multiple generations.

Speaking of massive replication of information, privacy seems to be in short supply nowadays. Cheap, easy-to-use video cameras and cheap data storage lead to the temptation to record everything happening in public or even private places and save it forever, with ensuing loss of privacy and potential loss of liberty, if despotic rulers get control of the data. But these recordings are sometimes good, deterring governmental as well as individual misconduct. In many situations, the bad guys want privacy for their misdeeds, while the good guys want publicity with authenticity.

To the amazement of most of the rest of the world, some Americans think it is good for society for everyone to carry a gun. A better idea would be for everyone to carry a camera. Public policy would then encourage amateurs to make audiovisual recordings, but restrict how the recordings could legally be used (yes to exposing crime and injustice; no to blackmail). A few years ago, I noticed a CNN billboard in Delhi with the slogan: *If you see it, shoot it—Every citizen a photojournalist.*

Returning to science, it seems there are three levels of privacy:

Quantum: Information, like the path of an unobserved photon, that exists only temporarily, and afterward can best be thought of as never having existed.

Classically Private: Information that has been amplified to the point of becoming classical, but is not widely distributed in easily recoverable form. Humans can erase it, then lie about it with impunity, although perhaps not without guilt.

Public: Information that is so widely distributed that it is infeasible to conceal. Lying about it only makes you look foolish.

Nowadays, it is tempting to believe that once information has become public and entered the blogosphere, it can never be wholly destroyed. The modern world appears very different in this regard from the ancient pre-Gutenberg era, when major literary works were written down, performed and widely known, but then lost.

In China, *The Classic of Music*, or 'Sixth Classic', is thought to have been lost in the book burning instigated by Emperor Qin Shi Huang in the 3rd century BC, though some general knowledge about it survives. Fortunately, Confucian scholars had memorised, and later managed to reconstruct, many of the other destroyed works. Sappho's poems were lost more gradually, through neglect: once widely reproduced and taught, they fell out of favour when her Aeolian dialect of Greek died out. They were no longer taught, and the existing manuscripts were discarded or repurposed. More recently, after surviving for over 1,000 years in India, the *Carvaka* school of philosophy is thought to have died out around the 15th century, along with all its original texts, except for fragments quoted in the writings of its Hindu and Buddhist opponents, who disliked it because of its denial of the afterlife, reincarnation and gods.

Even in today's world, much macroscopic, publicly accessible information is seemingly lost because no person, nor any natural process, happens to record it in a durable medium. For example, I took a photograph of craters made by raindrops in mud in a riverbed in Las Vegas, USA, in 1965. A few days later, these details were washed away by a subsequent heavy rain. If no one had photographed them, would a physical record of them still exist? It is tempting to believe that such macroscopic information is not really lost, just that it becomes so diffusely and complexly spread out as to be irrecoverable in practice while being still recoverable in principle. When a book is burned, its contents are in principle still recoverable from the exact state of the smoke, ash and heat it generates. Could it be that every macroscopic past phenomenon, say Sappho's lost poems, or the fate of mysteriously disappeared persons, like US labour leader Jimmy Hoffa, can be recovered from physical evidence in principle, if not in practice? To believe otherwise is venturing dangerously close to the postmodernist view, abhorred by most scientists as arrogantly anthropocentric, that the past (or maybe even the present) has no objective reality independent of human belief systems, and therefore that it is pointless to inquire what 'actually' happened.

But I think some information really is lost, not from the universe, but from the world (i.e. the planet Earth). Why? Because most information we might care about is washed away by much larger entropy flows into and out of the Earth. The Earth has finite information storage capacity, but it exports a lot of randomness (generates a lot of entanglement with its environment, in the quantum way of speaking) in the form of thermal radiation into the sky. The thermal entropy export rate ≈ 300 watts/sq meter at 300K ≈ 10^{30} bits per square meter per year.

The geological information capture rate in 'hard' degrees of freedom, stable for geological times against thermal motion (e.g., atomic substitutional disorder and crystal lattice defects in the solid rock of the Earth's crust) can be estimated as crust thickness (≈ 10 km) × rock information density (≈ 1 bit/cubic nm)/ rock lifetime (≈ 10^8 yr), or about 10^{22} bits / per square meter per year, which is far too low to remember everything that happened.

To catch up with the radiation leaving Earth, one would need to travel faster than light. One might say that the information is still in the universe, but not recoverable by us. So we are motivated to add a new level of privacy.

Quantum: Information, like the path taken in an interferometer, exists only temporarily, and afterward can best be thought of as never having existed.

Classical but Escaped: Information that has been amplified to the point of becoming classical, but has escaped from Earth in thermal radiation. Humans have no way of recovering it.

Classically Private: Information that has been amplified to the point of becoming classical and still resides on Earth in a few places, though it may be infeasible to recover with current technology.

Public and Permanent: Information that is so widely distributed that it is infeasible to erase all the copies.

Mysteries of the past

Some information about past events is still recorded on earth, though unknown to any human and inaccessible with current technology; for example, locations of gold rings, dropped in an annual ceremony into the Venice Lagoon over a period of several centuries, to symbolise Venice's marriage to the Sea.

Other information *may* be still recorded on Earth or may have escaped. Examples include the lost classic writings of many cultures, the fates of mysteriously disappeared persons such as physicist Ettore Majorana, who disappeared in 1938, labour leader Jimmy Hoffa, who disappeared in 1975, and computer scientist Jim Gray, who disappeared in 2007, or unrecorded raindrops from past rainstorms.

The perfect crime

How could one obliterate earthly evidence of Jimmy Hoffa's demise? (The former US labour leader disappeared in 1975, presumably murdered by the New York City Mafia, but his body was never found. Police are still searching.) What would one do to make sure the information of how and where he was killed could never be discovered? We suggest the following:

(a) Cremate his body in an out-of-sight location and let the smoke and heat escape.
(b) Dissolve the ashes to make a clear liquid with no solid fragments, then pour the liquid into the ocean.
(c) Don't tell anyone you did it, even on your deathbed.
(d) For good measure, have yourself cremated and your ashes dissolved to make sure physical traces of your memory are thoroughly gone.

But is radiation leaving the Earth really unrecoverable? Can we arrange for escaped information to be reflected back to us later, making it again accessible? Yes, for specific items of non-thermalised outgoing radiation (e.g. optical earth views, old TV broadcasts). This could be arranged, with advance planning, or it might happen accidentally. Such information could be called *extraterrestrial fossils.*

But for fully thermalised radiation, one would have to catch and reflect back so much of the radiation leaving Earth to reconstruct any particular item of interest that the Earth would have a serious climate change problem.

Randomising dynamics in a representative case

Though a raindrop originates in quantum and thermal fluctuations, it does not fall in a superposition of places. Independent observers would agree where it fell, and as long as the drop or its crater exists, reflected light will generate a torrent of replicas of the information that escape into space. However, unless the crater is lucky enough to get fossilised, it will be washed away, and its former location will then lose any stable earthly embodiment. The torrent of optical replicas will cease, and the old optical replicas will escape into space. So it would appear that the classical information, of where it formerly was, remains in the universe, but not on Earth.

Ontological status of escaped information

Consider a raindrop that may fall in one of two terrestrial locations **L** or **R**. Suppose that it forms, falls and finally evaporates, so that all earthly record of where it fell is lost as radiation into the sky. The possible states are:

(**LLL**_L_ + **RRR**_R_)/√2: A drop forms, falls and begins to emit radiative replicas (italics) into space. All observers, terrestrial and celestial, will see the drop as having fallen in one of two places. God sees a cat state-like superposition in which both outcomes happen.

(**LL**_LLL_ + **RR**_RRR_)/√2: Drop begins to evaporate, emitting further radiative replicas.

(_LLLLL_ + _RRRRR_)/√2: Drop has entirely evaporated. No terrestrial information remains where it fell.

Conclusion: escape of last replica from Earth restores terrestrial observers to a more detached, Olympian viewpoint in which both outcomes are equally real. Escaped information is not so different, after all, from which-path information. According to J. A. Wheeler, ‘The past exists only insofar as it is recorded in the present’.

Most classical information, such as the pattern of snowflakes on the ground last winter, is impermanent, eventually losing its durable embodiment and escaping from the Earth in outgoing radiation. Occasionally, information is lucky enough to get fossilised by natural processes or recorded by humans in a durable medium. Such information can last billions of years.

Escaped information still exists in the universe, but it is inaccessible on Earth. We have little justification for continuing to think that one alternative really happened but the others did not.

Note that even though I have argued that escaped information no longer has a preferred value, it still has a preferred *basis,* according to quantum Darwinism. One form of the Copenhagen interpretation (presuming a unitarily evolving Earth but an irreversible measurement process somewhere in the sky) says that escaped information does have a definite value, which we are ignorant of. If we find an extraterrestrial fossil, it will ‘agree’ with the value we once knew but have forgotten. Ruediger Schack says he would rather believe that Sappho’s lost poems are real than that the wave function of the universe is real. I lean the other way, but it is only a matter of taste.

Complexity

Enough about information and remembering and forgetting. Can we find a non-anthropocentric definition of what kind of information is *worth* remembering? How should *complexity* be defined? What is its connection with the universe not being at thermal equilibrium?

A simple cause can have a complicated effect, but not right away. Self-organisation is the spontaneous increase of complexity. A simple dynamics (e.g., the reversible deterministic cellular automaton illustrated below) can produce a complicated effect from a simple cause. (This automaton is a two-state second-order deterministic Ising model, where the future depends on the previous two time slices, the future of each cell differing from its past if and only if exactly two of the four nearest neighbors, not counting the cell itself, are black in the present time slice.) A small irregularity in initial configuration at the left edge (time proceeds horizontally) produces a complex deterministic 'wake' spreading out behind it. A sufficiently big piece of the wake contains enough evidence to infer the whole history. A smaller piece does not.

In the philosophy of science, the principle of *Occam's Razor* directs us to favour the most economical set of assumptions able to explain a given body of observational data. The most economical hypothesis is preferred, even if the deductive path connecting it to the phenomena it explains is long and complicated. In a computerised version of Occam's Razor, the hypotheses are replaced by alternative programmes for a universal

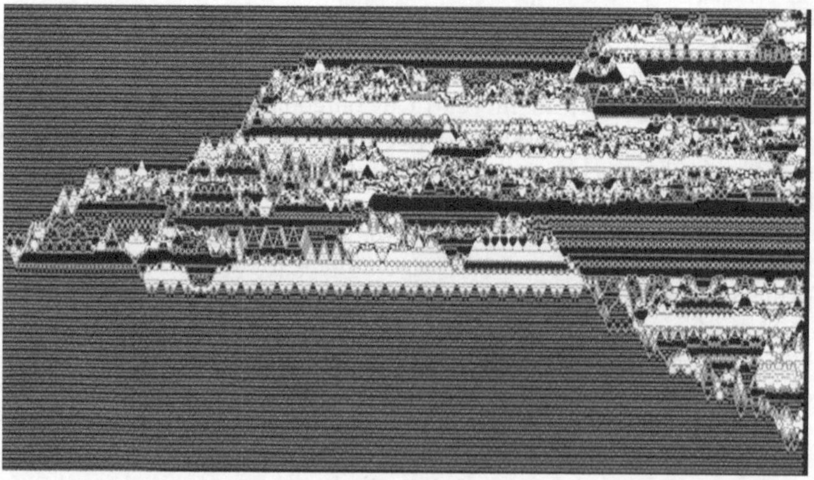

Figure 3.15

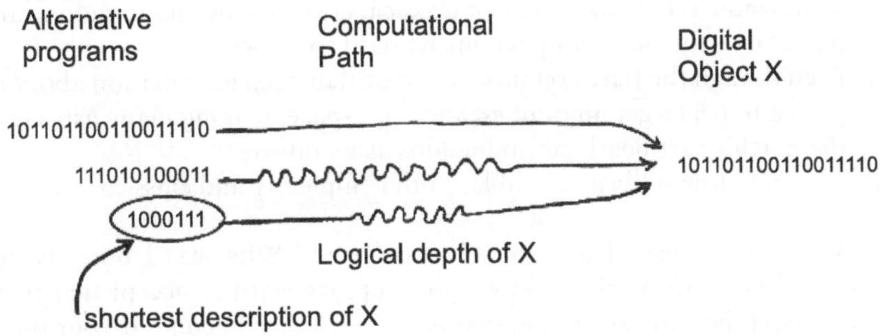

Figure 3.16

computer to compute a particular digital or digitised object **X**. The shortest programme is the most plausible, so its *run time* measures the object's *logical depth*, or the plausible amount of computational work required to create the object. A trivially orderly sequence like 111111. . . is logically shallow because it can be computed rapidly from a short description. A typical random sequence produced by coin tossing is also logically shallow, because it is essentially *its own* shortest description, and it is rapidly computable from that. Depth thus differs from Kolmogorov complexity or algorithmic information, defined as the *size* of the shortest description, which is high for random sequences.

If a reversible local dynamics (e.g. the one-dimensional system considered earlier) is allowed to run long enough in a closed system, the state becomes trivial and random, a discrete version of 'heat death' in thermodynamics. Our world is complex only because it is still out of equilibrium. After equilibration, a typical time slice is shallow, with only local correlations.

In an equilibrium world with local interactions (e.g. a thermal ensemble under a local Hamiltonian), correlations are generically local, mediated through the present. By contrast, in a nonequilibrium world, local dynamics can generically give rise to long-range correlations, mediated not through the present, but through a V-shaped path in space-time representing a common history.

Conclusions

In place of Laplacian determinism, quantum mechanics gives us a world where

(a) Many aspects of the future are inherently ambiguous: Even God does not know which radioactive atoms will decay, or who will win next year's elections. It is unreasonable to want to know some of these things beforehand.

(b) In a world out of thermal equilibrium, the monogamy of entanglement leads to the emergence of classical correlations and paradoxically makes overtly quantum phenomena hard to notice.
(c) Even though the Earth retains a great deal of deep information about its past, a much larger amount escapes into space, making many aspects of the Earth's past nearly as ambiguous as its future.
(d) Thermal disequilibrium enables both complexity and classicality.

As children, most of us probably wondered, 'Why was I born as me, instead of as someone else?' As we grow up, we learn to accept that there is no answer, because swapping two people—so that I acquired your body, your memories etc. and you mine—would make no difference in the state of the world as seen by an omniscient God. Quantum mechanics teaches us that the distinction between what happens and what might have happened can be regarded as this kind of distinction without a difference. This is the Everett picture. Alternatively (and equivalently, for scientific purposes), the distinction may be viewed as purely stochastic, with no underlying cause or explanation. That is the Copenhagen picture.

More philosophically, to what extent should science seek to economically describe the world as it 'is'—as it appears to us? Or is it sufficient for it to describe an ensemble of possible worlds, within which the one we see is typical, either absolutely, or relative to what sentient observers would see? How does one go about thinking about other possible worlds, or other kinds of sentient observers, without leaving out most of them through insufficient imagination? Scientists are generally leery of such anthropic reasoning because it is hard to think anthropically without thinking anthropocentrically and departing from the remit of science to make testable predictions. (My late IBM colleague Phil Seiden said scientists on Jupiter routinely discount the possibility of life on Earth because complex organisms could not possibly survive in its corrosive near-vacuum atmosphere.) But it is unreasonable to deny that some sort of anthropic selection may be biasing our view of the universe.

A scientific approach is to resort to anthropic reasoning as little as possible (weak anthropic principles) and attempt to use it only in situations where some testable predictions can be made or some reasonable conclusions drawn, albeit with low statistical confidence, e.g.,

(a) Early origin of life on Earth together with the relative recentness of the Cambrian explosion (weakly) suggests that multicellularity is improbable compared to life, but conditioned on multicellularity, the emergence of intelligence is likely.
(b) The failure to detect extraterrestrial intelligence (Fermi Paradox) suggests that the density of civilisations able to communicate over intergalactic distances is lower than the density of galaxies.

Returning to the Einstein-Tagore debate over the extent to which reality is influenced by humanity, various cosmologists have suggested that seeming accidents such as the small-but-nonzero cosmological constant or the anisotropy of the cosmic microwave background may arise from a Schopenhauer form of the weak anthropic principle: We live in the worst of all possible worlds, in other words, a world just barely compatible with our existence (or the existence of something complicated, maybe just Science).

Appendix: the mathematics of quantum mechanics

To each physical system, there corresponds a Hilbert space of dimensionality equal to the system's maximum number of reliably distinguishable states. A Hilbert space is a linear vector space with complex coefficients and inner product (where * denotes the complex conjugate):

$$<\varphi,\psi> = \Sigma_j \, \varphi_j^* \, \psi_j$$

(a) Each direction (ray) in the Hilbert space corresponds to a possible state of the system.
(b) Two states are reliably distinguishable if and only if their Hilbert space vectors are orthogonal. For example, horizontally and vertically polarised states can be represented by these orthogonal vectors in two-dimensional space.

Figure 3.17

But any other vector in this two-dimensional space, with real or complex coefficients, is also a possible state, for example, the two-diagonal polarisations:

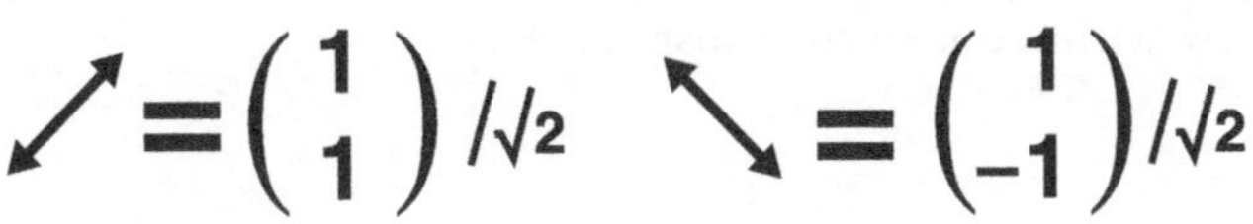

Figure 3.18

(Division by the square root of 2 keeps the vectors of unit length, the standard convention.)

(c) Spontaneous evolution of an unobserved system corresponds to unitary (i.e. orthogonality-preserving) evolution of its Hilbert space vector. Unitary evolution is the complex, multi-dimensional generalisation of a rigid rotation in three-dimensional space.

(d) The Hilbert space of a composite system is the tensor product of the Hilbert spaces of its parts. Thus, the state space of two separate polarised photons is four-dimensional, allowing them to exist in so-called **product states** such as

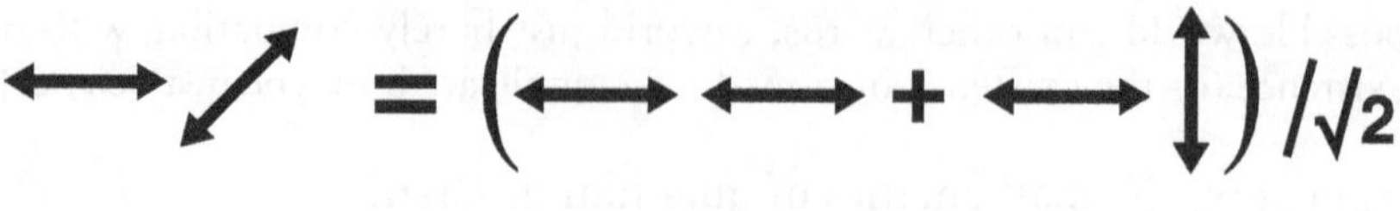

Figure 3.19

where the first photon is horizontal and the second diagonal, and also in so-called **entangled** states such as

Figure 3.20

in which neither photon has a definite state, even though the pair together does.

(e) Each possible measurement on a system corresponds to a resolution of its Hilbert space into orthogonal subspaces $\{P_j\}$, where $\Sigma_j P_j = 1$. On state ψ, the result j occurs with probability $|P_j \psi|^2$, where $|.|$ denotes the length of a vector, and the state after measurement is $P_j |\psi\rangle / |P_j |\psi\rangle |$. Believers in the 'many worlds interpretation' reject this axiom as ugly and unnecessary. For them, measurement is just a unitary evolution producing an entangled state of the system and measuring apparatus. For others, measurement causes the system to behave probabilistically and forget its pre-measurement state, unless that state happens to lie entirely within one of the subspaces P_j.

Notes

1 Wiesner, S. J., 'Conjugate coding', *Sigact News* 15 (1983), pp. 78–88; original article written circa 1970.

2 Bennett, C. H. and Brassard, G., 'Quantum cryptography: Public key distribution and coin tossing', in *Proceedings of IEEE International Conference on Computers, Systems and Signal Processing*, December 1984 (New York: IEEE, Bangalore, India, 1984), pp. 175–79.

Ekert, A. K., 'Quantum cryptography based on Bell's theorem', *Physical Review Letters* 67(6) (1991), pp. 661–63.

3 Einstein, A., Podolsky, B. and Rosen, N., 'Can quantum mechanical description of physical reality be considered complete?', *Physical Reviews* 47 (1935), p. 777. See also

Bell, J. S., 'On the Einstein Podolsky Rosen paradox', *Physics* 1 (1964), pp. 195–200.

4 Hughston, L., Jozsa, R. and Wootters, W., 'A complete classification of quantum ensembles having a given density matrix', *Physics Letters A* 183 (1993), pp. 14–8.

Hadjisavvas, N., 'Distance between states and statistical inference in quantum theory', *Annales de l'Institut Henri Poincar´e–Section A* 35(4) (1981), pp. 287–309.

Gisin, N., 'Comments on 'Assumptions implying the Schrödinger equation', by Thomas F. Jordan *American Journal of Physics* 59 (1991), pp. 606–8, *American Journal of Physics* 61(1) (1993), pp. 86–7.

5 Blume-Kohout, R. and Zurek, W. H., 'Quantum Darwinism: Entanglement, branches, and the emergent classicality of redundantly stored quantum information', *quant-ph/0505031 v2* (October 2005).

Riedel, C. J. and Zurek, W. H., 'Quantum Darwinism in an everyday environment: Huge redundancy in scattered photons', *arXiv:1001.3419v3* (July 2010).

4 Einstein and Tagore, Newton and Blake, Everett and Bohr

The dual nature of reality

Anthony Sudbery

In their conversation on the nature of reality on 14th July, 1930,[1] Albert Einstein and Rabindranath Tagore personified the division between two broad classes of attitude toward the world: one matter-based, centred on an external world in which human beings form a small part and emphasising the thinginess of things, the other mind-based, centred on human experience and emphasising human creativity and imagination. This characterisation may suggest a division between the scientific and artistic sensibilities, and those are indeed the roles played by Einstein and Tagore in their dialogue; but proponents of the two worldviews do not separate so neatly according to their pursuits. Moreover, just as scientific and artistic interests are not necessarily antagonistic, so these two attitudes to reality are not necessarily contradictory. I will argue that the effort to achieve a fully scientific understanding of the world impels us to adopt both stances simultaneously.

For Tagore, everything relates to the human mind: '[The] Truth of the Universe is human Truth'.[2] Einstein, on the other hand, believes in 'the world as a reality independent of the human factor'.[3] These seem appropriate positions for the poet and musician on the one hand, working always with immediately felt experience, and the scientist on the other, seeking to detach himself from his fallible senses and attain impersonal certainty by the light of cold reason. Interestingly, though, it is Tagore who attempts to support his position by rational argument, while Einstein simply states his position as a matter of faith: 'I cannot prove that my conception is right, but that is my religion'.[4] His part in the conversation concludes, 'Then I am more religious than you are!'

Tagore argues that

> This world is a human world—the scientific view of it is also that of the scientific man. Therefore the world apart from us does not exist.[5]

This is reminiscent of Jacques Derrida's famous (if mistranslated) pronouncement 'There is nothing outside the text',[6] which inspired modern (or perhaps post-modern, i.e. pre-21st century) sociologists of scientific knowledge like Andrew Pickering, for whom all scientific knowledge is a 'social construction' and who see it as their task to explain what scientists believe without referring to the content of those beliefs. We do not believe in quarks because there really are quarks which scatter electrons in certain ways, but because . . . well, I am not quite sure why, but it is all very sociological.[7] Pickering believes that the scientist's explanation of our belief in quarks is circular, because he can see no difference between 'Quarks exist' and 'We believe that quarks exist'. Many scientists will find Tagore's view as uncongenial as Einstein did, but well-brought-up quantum physicists may have qualms. Didn't Bohr teach us that there is no world beyond the scale of laboratory apparatus that we have constructed? Doesn't this support Tagore's view, at least in the form 'the world apart from our apparatus does not exist'?

Einstein felt no need to give Tagore arguments in support of his belief in an external reality. He presents it as a matter of faith. It could perhaps be seen as a statement of intent: 'I am determined to understand the world as an external reality'.[8] In these terms, he might have seen Bohr's quantum mechanics as an admission of defeat. Henry Margenau commented,

> Like most scientists, Einstein leaves unanswered the basic metaphysical problem underlying all science, the meaning of externality,[9]

but if belief in external reality is understood as a statement of intent, then the belief needs no justification and 'externality' needs no definition; Einstein can simply say 'I will know it when I see it' and go on looking for such an understanding.

If one cannot completely analyse the concept of 'external reality', one can certainly point to particular theories that exemplify it. The paradigmatic example is Newton's dream of a purely mechanical theory:

> I wish we could derive the rest of the phenomena of Nature by the same kind of reasoning from mechanical principles, for I am induced by many reasons to suspect that they may all depend upon certain forces by which the particles of bodies, by some causes hitherto unknown, are either mutually impelled towards one another, and cohere in regular figures, or are repelled and recede from one another.[10]

This has inspired the view that physical reality consists entirely of point particles, each of which has a definite position at every instant of time, together with various other numerical properties such as mass and electric charge; the motion of these particles is completely determined by the forces between them. The reality of these particles owes nothing to human minds; it seems to me to be a prime example of what could be meant by a concept of an objective, external reality.

A statement that the physical aspects of reality could be exhaustively described in such a way is not necessarily a denial that there are also human aspects to reality, such as creativity, free will, morality and so on. But it can certainly look like such a denial, and therefore arouses hostility in a romantic, artistic sensibility. This hostility is seen above all in William Blake. Blake exalted freedom and creativity and regarded reason as something to be fought:

> May God us keep
> From single vision and Newtons sleep.[11]

In his mythology, the restrictive, tyrannical God (and creator of the physical world) is called Urizen, which can be heard as 'your reason'.

Blake's famous picture of Newton[12] shows this single vision. Newton's objective reality is represented by a mathematical diagram, pale and *un*real compared to the rich, colourful, but subjective reality which he is prevented from seeing by his concentration on the objective but abstract, mathematical aspects of reality. Blake does not deny the validity of Newton's vision; it is notable that Newton's mathematical instrument is the same as that held by God (Urizen) in *The Ancient of Days*,[13] Blake's picture of the creation of the world. But he is hostile to the *singleness* of this vision. In the words that Einstein used about quantum theory, he denies its completeness.

Einstein's idea of reality had different contents from Newton's, but I think he shared with Newton the idea of reality itself. They differed as to what kinds of thing are real, but they agreed that there was something real, whether or not it is observed or described, and independent of any observation or description. Einstein's commitment to realism, in this sense, was a major reason, though not the only one, for his resistance to the claims of quantum theory to be a possible final theory of the world. It is explicitly adduced as the ground for his famous challenge in the Einstein, Podolsky and Rosen (EPR) paper[14] to the completeness of 'the quantum-mechanical description of reality'. This argument also requires a principle of locality—As he later stated it, 'the real states of spatially separated objects are independent of each other'.[15] But this is introduced in an almost casual, incidental way at the end of the EPR paper; it is the concept of reality that bears all the weight of the argument. And this was only to be expected; as it had taken shape as a coherent conceptual structure in the hands of Bohr and Heisenberg, the understanding of quantum theory had come to be seen as inseparably linked with an anti-realist attitude to scientific knowledge. Bohr repeatedly insisted that the equations and mathematical objects of the theory were not to be seen as pictures of a microscopic physical reality, but as algorithms for calculating macroscopic effects that are all that humans can presume to know—a position that was rejected by Einstein, Podolsky and Rosen from the outset, in the very words of their title. In 1935, EPR and Bohr both wrote papers entitled, 'Can the quantum-mechanical description of reality be considered complete?' EPR's answer was 'No'. Bohr's answer, essentially, was 'What reality?'

It is now generally agreed, following the brilliantly clear light that John Bell shone on the problem, that realism and locality together are indeed incompatible with quantum theory and that the incompatibility is not a metaphysical matter of opposing concepts, but a clear-cut question of conflicting predictions for the results of experiment. It has been possible to put the question to the test, and the experimental verdict seems (almost, but not quite, inarguably) a clear vindication of quantum theory.

Unfortunately, Einstein did not live to see Bell's theorem. He would no doubt have been as surprised as everyone else to see that the possibility of completing quantum mechanics could be tested so decisively without any need to consider what form the more complete theory might take. No doubt, also, he would have felt the experimental results as a jarring blow to his worldview. But, perhaps surprisingly, he would not have had to abandon his fundamental faith in the existence of external physical reality. There are in fact two ways in which quantum theory can still be seen as a realist description of the world. Both would have been found uncongenial by Einstein, but perhaps not as uncongenial as abandoning his faith in external reality.

The first realist conception of quantum theory must have been known to Einstein, as it was proposed by Louis de Broglie at the 1927 Solvay conference at which Einstein was present. In de Broglie's theory, realism is achieved in just the way assumed in the EPR paper: the quantum description of a collection of particles is acknowledged to be incomplete and is completed by further elements of reality in the form of precise positions for each of the particles. However, it explicitly violates the principle of locality, which was also assumed by EPR. At the Solvay conference, it was subject to vigorous attack, in particular by Pauli, in the face of which de Broglie withdrew the theory, though it was later revived by David Bohm, who showed that at least some of Pauli's criticisms were unfair. The nonlocality is undeniable, however, and makes it difficult to reconcile this theory with relativity. Even now that nonlocality is accepted by many physicists as an actual feature of the world, there is little support for the de Broglie-Bohm theory.

The second option for realists only emerged after Einstein's death, in 1957, when Everett and Wheeler published their 'relative-state interpretation' of quantum theory.[16]

Here we have an opposite view to EPR: instead of regarding it as incomplete and in need of further elements, Everett and Wheeler take the quantum-mechanical description of reality completely seriously and indeed, remove from it an element that had always seemed artificial and awkward. This element, known as the 'collapse of the wave function', was probably motivated by realist hankerings and was not found necessary by those who adhered rigorously to Bohr's anti-realist doctrine, but it sits uneasily with any attempt to take quantum theory as a realistic account of the world. However, if the collapse of the wave function is removed from this account, the world—or rather, according to later accounts, many worlds—that emerges is so strange that at first few people could accept it. In order to discuss this theory and defend my characterisation of it as a realist view, I must step back and describe the orthodox quantum theory from which it arose.

The following features of quantum theory had become established as common ground by 1935. Associated with any physical system is a mathematical object (its *wave function* or *state vector*) that changes in time according to a well-defined law (the Schrödinger equation) determined by the forces acting on the system. The natural tendency is to regard this as analogous to Newtonian mechanics, in which the physical system is a collection of particles, the mathematical object is a collection of geometrical points and velocity vectors and the change in time of this object, determined by the forces acting on the particles, is given by Newton's laws of motion. However, in Newtonian mechanics, one can find out the exact mathematical description of the system by observing the positions and velocities of the particles; conversely, a specification of the mathematical description tells one exactly what one would find on observing the system. In the microscopic world of quantum theory, by contrast, it is not possible to observe enough properties of the system to determine its state vector (one sometimes knows the state vector if one has prepared the system in a particular way, but it is not possible to find out how someone else has prepared the system just by observing it), and the mathematical description does not, in general, tell you what you can expect to see on observing the system. It only tells you the possible results of an observation and the *probabilities* of these different results.

This state vector is not usually easy to see as a picture of the system, in the way that the Newtonian mathematical object is a picture of a set of particles in definite positions moving in definite directions with definite velocities. There is indeed a form of the state vector that corresponds to the particles being in particular positions (though not velocities), but there is another form corresponding to a different configuration of the particles, and there is a third form (the *sum* of the other two) that somehow contains both configurations. This sum tells us probabilities of finding *either* of the two configurations when we observe the system, and it is tempting to think of it as containing the information that the system is in *either* the one configuration *or* the other. But this will not do; its logical consequences are definitely wrong (it wipes out the 'interference effects', which show the wavelike nature of matter—one of the strange empirical findings that showed the necessity for quantum theory in the first place). As a result, it has become common to say that the third (sum) form of the state vector describes a system that is in *both* the first configuration *and* the second. The mathematical apparatus of vectors and vector sums seems to support this: The vector sum of a north-pointing velocity vector and an east-pointing velocity vector is a velocity vector pointing NE, describing motion in which one is *both* travelling north *and* travelling east. But it is certainly hard to picture an object that is both here and there, and it is clear why Bohr held that this mathematical structure should not be seen as any kind of description of a real object. 'There is no quantum world', he said. 'There is only an abstract physical description. It is wrong to think that the task of physics is to find out how nature *is*. Physics concerns what we can *say* about nature'.[17]

The situation is even worse than this: if the system in question is not isolated but is part of a larger system (as every actual system is part of the universe), then, in general, the system has no definite state vector. The larger system may be described by a state vector, but this cannot usually be analysed into unique descriptions of its parts separately. Instead, the overall state vector will be the sum of a number of components, each of which describes one of the parts in a definite state and the other part in an associated state. The two parts are said to be *entangled*.

I have spoken of 'observing' the properties of a system, which is appropriate language for a realist; there is the physical system, out there, with its properties, and the physicist, separate and detached, quietly observing. Every student of physics is taught that this is wrong, that the properties can only be discovered by *measuring* the system in an active experimental intervention, which, on the scale on which quantum theory is relevant, will inevitably alter the very properties one wants to observe. For this reason, my account of quantum theory would be frowned at by most physics teachers: wherever I have used the word 'observation', it should be replaced by 'measurement'. This word, indeed, has become an important and fundamental part of the theory, with the character of a primitive undefined term in the fundamental postulates. In its most austere (and anti-realist) form, the theory renounces any pretension to describe a physical world; it only offers a method for calculating the results of laboratory procedures, each consisting of a specified preparation followed, after a specified lapse of time, by a specified measurement. The calculation will tell us the possible results of this measurement and the probabilities for each of them.

This is neither satisfying nor satisfactory. It is not satisfying because, to many scientists, it misses the whole point of doing science in the first place. Why would we want to predict the results of experiments? We want to understand the world, and we do experiments to check if we've got it right.[18] As John Bell put it,

> To restrict quantum mechanics to be exclusively about piddling laboratory operations is to betray the great enterprise.[19]

It is not satisfactory because it is hardly possible to obey the stern injunction to consider only laboratory procedures consisting of a single preparation followed by a single measurement. Often one wants to follow the progress of events beyond the final measurement. In that case, one has to treat the measurement as a second preparation, giving a new state vector for the system which will serve for the start of a new calculation. But the nature of this preparation, and therefore the identity of the new state vector, depends on the result of the experiment, which is a random outcome over which the experimenter has no control. If we follow the state vector after the first preparation, through the measurement and beyond, we see an evolution that changes abruptly and randomly at the time of the measurement. This is incorporated in the presentation of quantum mechanics in most textbooks, which (following the original formulation of Paul Dirac[20] and John von Neumann)[21] stipulates that the state vector changes in two distinct ways:

1 Left to itself, the system changes smoothly according to the Schrödinger equation; but
2 If the system is subjected to a measurement, its state vector changes instantaneously and unpredictably to reflect the result of the measurement. It undergoes *projection.*

Physicists have always felt uneasy about this. It is supposed to be a fundamental statement of a basic law of nature, on a par with Newton's laws of motion. But it lacks the unity and simplicity of Newton's laws: why are there two quite different laws applying in different situations? And anyway, what are these different situations? What is this 'measurement' that enters into the second law as an undefined term? Surely measurements are not basic constituents of the world. An actual measurement in a laboratory is conducted with apparatus that can be analysed as a physical system like any other, and must itself follow the Schrödinger equation as in the first postulate.

This is the measurement problem of quantum mechanics. Von Neumann investigated it by taking the measurement seriously as a physical process and considering the quantum description of a total system consisting of the object being measured together with the measuring apparatus, which interact by means of known forces between them. Following the progress of the measurement according to the Schrödinger equation yields a total state in which the object and the apparatus are entangled: the total state vector is a sum of states in each of which the apparatus shows a definite result and the object definitely has the property shown by that result. But this total state vector does not correspond to any particular result. In order to reflect the fact of experience that a measurement does have a definite unique result, von Neumann found it necessary to appeal to the second (projection) postulate after all.

At this level, it becomes very tempting to interpret the quantum sum of state vectors in terms of 'or' rather than 'and'. In the laboratory it is surely true, after a measurement has been made, that it had one result *or* another. I said earlier that this interpretation leads to conflict with experimental facts, in that it predicts that there will be none of the interference effects which are characteristic of quantum phenomena (and which are predicted by the Schrödinger equation). But for large things like laboratory apparatus, these interference effects are far too small and infrequent to be ever observed; it will never lead to a conflict with an experiment to assume that the quantum A + B means that either A happened or B happened. This result is known as *decoherence*; it is well established in theory.[22] Nevertheless, it remains an approximate statement; in principle, the exact statement about the apparatus and the object is that their state vector is the sum of different results, and this cannot be understood as stating that some one of these results happened.

In 1957, Hugh Everett challenged von Neumann's conclusion. Why can't we believe, he asked, that the world is described by a state vector which is the sum of components describing situations which we would recognise as different and incompatible states of affairs? The immediate answer is that the world just isn't like that—we see that it isn't; we never see such sums. In terms of Schrödinger's famous example,[23] we never see a cat in a state that is the sum of being alive and being dead. Everett's reply (anticipated by Schrödinger himself) was that the theory *tells* us that we will never see such a cat. Since we are ourselves physical objects, the universal state vector must describe our brains as well as everything else. If I look at Schrödinger's cat after it has been in his diabolical box for a while, the physical process by which I see it leads, via Schrödinger's equation, to an entangled state that is the sum of two very different states: one in which the cat is dead and I see it as dead, and another in which it is alive and I see it as alive. Nowhere is there a state of my brain seeing a cat that is the sum of alive and dead. Everett made the analogy with Copernicus's revolutionary statement that the earth moves round the sum and rotates on its axis. To the objection that it doesn't *feel* as if we're whirling around in space, Copernicus could reply (though he may have had to wait for help from Galileo and Newton) that the laws of physics show that what it feels like to be whirling round in space (on a massive gravitating planet) is exactly what we do feel.

Following Everett, there has been a growing body of opinion among physicists in favour of taking quantum theory seriously and literally: it gives us a mathematical object (the state vector) that constitutes a complete description of the physical world. This state vector has many parts, each of which we can recognise as a picture of a world that we can understand—it contains atoms and molecules in combinations that we know as planets, and mountains, and trees and tigers and people. But different things are happening in each of these parts: in some of them, Schrödinger's cat is alive, in some it is dead, in some of them nobody has heard of Schrödinger's cat because the young Schrödinger decided to become a poet rather than a physicist. The theory contains 'many worlds'.

If this is right, our most successful physical theory gives us an account of reality that contains far more than what we see, or could ever see. If I do an experiment that has two possible outcomes, A or B, depending on whether a certain radioactive nucleus has decayed or not in the course of the experiment, then what I see is that the experiment went one way (say A) and not the other (B). I believe that A happened and not B (because I saw it); everyone I ask to check my experiment agrees that I am right. Surely, by all the standards of scientific truth, my belief is justified and I am entitled to say that in the *real* world, A happened and B did not. Yet the theory tells me that there is another world, also real, in which A did not happen but B did.

So why don't we look again at the meaning of the quantum description, with its many worlds? We can have a mathematical description containing all these worlds and believe that it is really true without believing that each of the worlds is real. Maybe the vector sum occurring in the mathematics should be interpreted as a disjunction: the statement is that all of these worlds are possible, but only one is real. The problem with this is that it requires a precise definition of what will count as a world, as a matter of basic principle, and this goes against the whole spirit of the theory. We can look at a particular state vector of the universe and say, 'Oh yes, this seems to have a lump here which looks like the kind of world we know, with people doing experiments and getting unique results, and there's another lump there with people getting different results, and over there is a smudge which doesn't look like anything much, but they are all combined together with this vector sum idea', but this interpretation will always be ad hoc, and we cannot give a general rule for recognising worlds that will apply in advance to any state vector. Those who think this is a problem call it the 'preferred basis problem' of Everett's theory.

There seem to be two conflicting accounts of reality here, both authorised by science. On the one hand, science is based on empirical evidence: what is revealed by careful experiment is real. On the other hand, it advances by taking its successful theories seriously: the theoretical entities of a fundamental theory are real. This tendency has often been resisted by the cautious—many physicists at the end of the nineteenth century refused to believe in the reality of atoms—but it seems to me that the general lesson of history is that reality favours the bold.

It is not, of course, the first time that a general scientific theory has seemed to be in conflict with intuition. I have already mentioned the conflict between the theory that the earth spins on its axis and the obvious fact that the sun goes round the earth once a day. Rutherford's discovery that atomic nuclei are so small that atoms are mainly empty space seems to falsify our perception that the things around us are solid. The lesson of relativity that physics takes place in space-time is said to show that there is no such thing as the passage of time. Both determinism and indeterminism are thought, not necessarily by different people, to make free will impossible.

Scientists often respond to these conflicts by saying that science has revealed that the intuitive belief is an 'illusion'. We have the illusion that the sun goes round the earth; the solidity of stones, the passage of time, free will—they are all illusions. In nearly every case, I think this is a mistake. It is not an illusion that the sun goes round the earth; Einstein taught us that we are free to adopt a frame of reference in which the earth is fixed (intuitively, we can hardly help doing so), and in that frame, the sun does indeed go round the earth. It is not an illusion that stones are solid, it is part of the meaning of solidity, and it is certainly not an illusion that we have free will, it is a clear fact of everyday life (once again, it is a matter of the meaning of words, though not an easy one to tease out in this case). In both these cases, the problem arises from confusing a clear, everyday concept with an unjustified pseudo-scientific theory of that concept: solidity does *not* mean that matter occupies a mathematical continuum, free will does *not* mean that we have the power to interrupt the laws of physics. As for the idea that time is an illusion, which was even held by Einstein, I have no idea what that is supposed to mean. (What is that we mistakenly believe when we are under this illusion?)

In all these cases, I believe the resolution of the conflict is not that one of the conflicting ideas is mistaken or illusory, but that there is no conflict. I would like to apply the same strategy to the conflict between the many worlds of quantum theory and the one world of experience, but it will involve some re-examination of our conception of reality.

I see this conflict as one of a wide class of philosophical problems that was identified by Thomas Nagel.[24] These problems arise whenever we attempt to go beyond our own individual situation and experience to obtain a more general, objective account of the world—-something that we may feel obliged to do for ethical, political or aesthetic as much as for scientific reasons. We want to lay aside our individual interests and work for the general good, to go beyond our subjective opinions and attain objective knowledge, to move out of our particular situation and see the overall picture, or, as Nagel puts it, attain a 'view from nowhere' as opposed to the 'view from now here'. The problem is that the objective propositions we reach, which we seem to have good reason to believe, then appear to conflict with the subjective experience which is still vividly present to us. Nagel acknowledges the force of the objective position, but rejects the temptation to dismiss the subjective experience as merely an illusion; the subjective experience has a vividness and a reality compared with which the objective truth is abstract, pale, 'etiolated'.

The contrast is beautifully depicted in Blake's picture of Newton. Newton is depicted as a thoroughly physical, muscular young man, but he is totally concentrated on the abstract mathematical laws of nature. He sees only the thin, colourless diagram on his scroll, and does not see the colour, texture and concrete reality of the coral-encrusted rock on which he sits. He sees all of reality; as we have already noted, the dividers he holds are the same instrument as that held by God in Blake's picture of the creation of the world. But he also misses all of reality, in the richness behind him.

(As I was writing this chapter, the radio was playing music chosen by an explorer. The music stopped, and I heard him explaining the purpose of his expeditions into the rainforest. 'I was not there as a scientist', he said, 'I was there to *understand* the place'. This paradox—are not 'science' and 'understanding' synonymous?—nicely captures the two sides of Blake's picture.)

Now, in the farthest we have travelled along the journey initiated by Newton, we find the reality of subjective experience being forced on us by the abstract theory itself. If we start as fundamentalist physicists, we have a quantum-mechanical state vector that we think describes the whole of physical reality at different times. Suppose we find that at a certain time, this state vector contains a world in which there is a scientist preparing Schrödinger's experiment, putting a cat in a box with a radioactive nucleus and a phial of volatile poison. At a later time, we will see that this part of the state vector has split into two: one part containing a dead cat and the scientist with his brain registering the thought 'Oh dear, the cat died', and another part containing a living cat and the scientist's brain registering the thought 'Ah, the cat is still alive'. From our external viewpoint, we see that neither of these thoughts is a true reflection of reality. But the thoughts do not occur in our external framework; each of them belongs in its own part of reality, and in that context it fulfils all the conditions for truth. It corresponds to an actual physical fact, it agrees with the thoughts of all other competent observers in that part of reality, and so on. Although the two thoughts are contradictory, each of them is true in its own branch of reality. It is like the truth of a sentence in a work of fiction:[25] the sentence takes its truth from the story in which it belongs. In St. Luke's Gospel, it is true that Christ died on the cross, but in Kazantzakis's novel *The Last Temptation of Christ*, it is true that he comes down from the cross and marries Mary Magdalene. A similar approach to truth and reality can be used to counter the sceptical thought 'Maybe none of this is real; maybe it is all a dream'. If it is a dream, then the thought 'This table is real' is occurring in the dream and as such is true: the table is real.

So there is one nature of reality for the many worlds described by quantum theory and another for the one world we actually experience. I like to think that the first view of reality reconciles quantum theory with Einstein's belief in a real external world independent of human observers (though I cannot be confident that Einstein would have agreed). The second (one-world) type of reality, however, explicitly depends on a particular observer for its definition. I believe that the phenomenon of decoherence makes it possible to replace this single observer with a community of observers, but still, the scientific definition of this genus of reality is linked to consciousness. Not necessarily human consciousness, in principle; nevertheless, this interpretation of quantum mechanics is remarkably close to Tagore's view. Tagore could have been referring to Everett's paper when he said (in a part of the conversation with Einstein which is not contained in Singer, ' "Endless dawns" of imagination', but was unearthed by Partha Ghose) '[the world] is a relative world, depending for its reality upon our consciousness.'

This is why my title has (three times) the word 'and' (rather than 'versus'). To understand quantum theory, we seem to need Einstein's reality *and* Tagore's, Newton's *and* Blake's, Everett's *and* Bohr's.

What are we, living in one definite world, to make of the quantum description of many worlds in a vector sum? How does it affect us? At the start of his experiment with the cat, Schrödinger knows the quantum state vector that describes how he has set up the experiment, and he can calculate what it will become at the end of the experiment. It will then describe a world in which the cat is alive and another in which it is dead. What does this calculation mean for Schrödinger (a) at the start of the experiment, when the result of the calculation lies in the future and (b) after the experiment, when it lies in the present?

1. Everybody agrees on how to use a quantum calculation. Before the experiment, the different worlds in the future state vector describe the different possible results. More than that, the relative sizes of these components give the probabilities for these different results.

2. Let us look on the bright side and suppose that the cat lives. Schrödinger has the immediate reality of the living cat in front of him; he also knows that there is another world with a dead cat in the quantum description. What can that mean, in his lived reality? He must take it as describing something that might have happened, but didn't. However, he can't cheerfully say, 'Oh good, it didn't happen, let's stop worrying about it' and start again with a state vector describing the living cat; that would be to ignore his own equation and apply the projection postulate, which gives wrong answers. The chance of an error is tiny, but in principle it is there. To calculate the results of future experiments, he must take into account the fact that the cat might have died. What might have happened, but didn't, can still affect the real world; therefore it is still a part of reality.

Einstein had two other independent objections to quantum theory. One of them was his belief in determinism. Realism and determinism are logically independent—a theory could exhibit either of them without the other—but it is not always clear that these were distinct for Einstein. In 1949, he was protesting

> I still believe in the possibility of a model of reality—that is to say, of a theory which represents things themselves and not merely the probability of their occurrence.[26]

In analysing this statement, one might note that it is not things that occur but events, and in asking for events themselves to be represented in the theory, without probabilities, Einstein seems to want all of space-time (past, present and future) to have a definite and univocal existence in the theory. Such a theory need not be deterministic, in the sense that identical causes must always be followed by identical effects, but it does require that the future is fixed (one might say it is 'fatalistic'). In any case, in requiring no probabilities, Einstein does seem to be asking for determinism in the strong sense. As is well known, he stated elsewhere that 'I, at any rate, am convinced that He [the 'old one'] does not throw dice'.[27]

In this respect also, Everettian quantum theory can be seen as meeting Einstein's desiderata (though, again, I am not sure that Einstein would have been happy with it). The objective, realistic aspect of the theory—the 'view from nowhere'—is fully deterministic, as the state vector of the universe obeys the Schrödinger equation. Given the state vector at any one time, the state vector at any future time (and, indeed, any past time) is completely determined. But, of course, the state vector relative to any given consciousness is subject to random changes; this aspect of the theory is completely indeterministic. The 'old one' does not play dice with the whole world, but he does play dice with each one of us. In the view from 'now here', the future is open;[28] the state vector of the universe gives us only a set of possibilities for our future, together with probabilities for each of them, but at any given time there is no such thing for us as *the* future. This makes it necessary to rethink what we mean, in this internal perspective, by probability and the truth of statements in the future tense.[29]

In the internal perspective, present truth and reality are guaranteed by my perceptions ('I' being the observer whose consciousness defines the perspective). Clearly, there can be no such guarantee from experience for the future. Nevertheless, it seems natural to believe that there is such a thing as my future experience, even though I do not know that there is a true statement about what my future state will be. This is denied by the quantum calculation: there is nothing to link any one of the components of the future state vector with the particular component that I am experiencing now. It might also be denied by a naïve view, perhaps our own view as children ('of course there isn't a definite future, it hasn't happened yet; anything could happen') and it was certainly denied by Aristotle, who, in his famous passage about the sea battle,[30] asserted that statements about the future are neither true nor false. This has been taken by many logicians as meaning that bivalent logic does not apply to future-tense statements; there is a third truth value as well as 'true' and 'false'. But Aristotle also pointed out that statements about the future can be more or less likely. This suggests that their truth value lies on a scale between 0 and 1 and should be identified with probability.

So from the internal perspective, there is one component of the present universal state vector that represents the truth now, and present-tense statements have truth values of 0 or 1 determined by their consonance with this component. Future-tense statements have truth values equal to the probabilities calculated by quantum mechanics.

This suggests further elaboration of the external perspective. In this perspective, the universal state vector represents the whole truth about reality. If this is the sum of many components describing worlds that we could recognise, then it is often said that these many worlds are 'all equally real'. But there is no warrant for that 'equally'. These components do not have equal lengths as vectors; if that length should happen to be zero, then the corresponding world is not real. A component with a length that is tiny but not actually zero, however, is supposed in this view to be fully real. It seems much better to say that full reality belongs only to the actual universal state vector; any other vector (for example, one of the components representing a recognizable 'world') has a lesser degree of reality measured by its contribution to the universal state vector. In symbols, if the universal state vector is

$$\Psi = c_1 \Psi_1 + c_2 \Psi_2 + \ldots$$

where $\Psi_1, \Psi_2, \ldots$ describe different recognisable worlds, then the degree of reality of the world Ψ_1 is $|c_1|^2$, that of Ψ_2 is $|c_2|^2$ and so on. More generally, if Φ is any other universal state vector, recognisable world or not, its degree of reality is $|\langle\Phi|\Psi\rangle|^2$.

In the external perspective, Everettian quantum mechanics is a realistic, deterministic theory—just what Einstein wanted. I feel pretty sure, however, that he would not have been satisfied with it. The problem is that neither the external perspective nor the internal one has a third feature that Einstein regarded as essential in a scientific theory, namely locality, or more generally, separability:

> Now it appears to me that one may speak of the real factual situation of the partial system S_2. . . . But on one supposition we should, in my opinion, absolutely hold fast; the real factual situation of the system S_2 is independent of what is done with the system S_1, which is spatially separated from the former.[31]

The existence of entanglement means that it is not possible to divide the world into pieces in such a way that the real factual situation of the whole world is completely described by putting together the real factual situations of its parts; indeed, the parts do not have their own real factual situations. In particular, one cannot divide space-time into regions and describe these regions independently of each other. The theory is *nonlocal* or *holistic*.

Conclusion

It has often been remarked that there is little engagement between Einstein and Tagore in their conversation on the nature of reality. They both state their views, which just sail past without affecting each other. The sole discussion between Everett and Bohr was described by Everett's biographer as 'simply a polite hearing and a lot of mumbling'[32] and by Everett himself as 'that was a hell of a—doomed from the beginning'.[33] It is hard to even imagine a meeting between Newton and Blake. My thesis here has been that although these incompatibilities are real, we do not have to choose one side or the other: We can understand the different contexts in which notions of truth and reality apply, and we can place ourselves in either context at will, without inconsistency. Quantum theory forces us to consider both, and each context shows us a different nature of reality.

Notes

1 Singer, W., '"Endless dawns" of imagination', *Kenyon Review* 23(2) (2001), pp. 7–33.
2 Ibid., p. 22.
3 Ibid.
4 Ibid., p. 23.
5 Ibid., p. 22.
6 '*Il n'y a pas de hors-texte*': http://en.wikipedia.org/wiki/Jacques_Derrida, accessed 28 March 2015.
7 A. Pickering, *Constructing Quarks* (Edinburgh: Edinburgh University Press, 1984).
8 For an account of Einstein's realism along these lines, see Fine, A., *The Shaky Game: Einstein, Realism and the Quantum Theory*, 2nd edn. (Chicago: University of Chicago Press, 1997).
9 Margenau, H., 'Einstein's conception of reality', in *Albert Einstein: Philosopher-Scientist*, P. A. Schilpp (ed.) (New York: Harper, 1949) pp. 245–68, on p. 249.
10 Newton, I., *Principia Mathematica* (Berkeley: University of California Press, 1934), Preface to the first edition, p. xviii.
11 Blake, W., 'Letter to Thomas Butts 22 November 1802', in *The Letters of William Blake*, G. Keynes (ed.) (Oxford: Oxford University Press, 1980), p. 46.
12 http://www.tate.org.uk/art/artworks/blake-newton-n05058
13 http://gallerysearch.ds.man.ac.uk/Detail/88
14 Einstein, A., Podolsky, B. and Rosen, N., 'Can quantum-mechanical description of physical reality be considered complete?', *Physical Review* 47 (1935), pp. 777–80.
15 Einstein, A., 'Reply to criticisms', in *Albert Einstein: Philosopher-Scientist*, Schilpp (ed.), pp. 665–88, on p. 682.
16 Everett, H., '"Relative state" formulation of quantum mechanics', *Reviews of Modern Physics* 29 (1957), pp. 141–53; Wheeler, J. A., 'Assessment of Everett's "relative state" formulation of quantum theory', *Reviews of Modern Physics* 29 (1957), pp. 463–65.
17 Quoted in Petersen, A., 'The philosophy of Niels Bohr', *Bulletin of the Atomic Scientists*, September 1963, pp. 8–14, on p. 12.
18 Sudbery, A., *Quantum Mechanics and the Particles of Nature* (Cambridge: Cambridge University Press, 1986) p. 214.
19 Bell, J. S., 'Against "measurement"', *Physics World*, August 1990, pp. 33–40, on p. 34.
20 Dirac, P. A. M., *The Principles of Quantum Mechanics* (Oxford: Oxford University Press, 1930).
21 von Neumann, J., *Mathematical Foundations of Quantum Mechanics* (Princeton, NJ: Princeton University Press, 1955). Translation by Robert T. Beyer of the original German, published in 1932.
22 Schlosshauer, M., *Decoherence and the Quantum-to-Classical Transition* (Berlin: Springer, 2007).
23 Schrödinger, E., 'Die gegenwärtige Situation in der Quantenmechanik', *Naturwissenschaften* 23 (1935), pp. 807–12, 823–28 and 844–49. Translated by J. D. Trimmer as 'The present situation in quantum mechanics: A translation of Schrödinger's "cat paradox" paper', in *Quantum Theory and Measurement*, J. A. Wheeler and W. H. Zurek (eds.) (Princeton, NJ: Princeton University Press, 1983) pp. 152–67.
24 Nagel, T., *The View from Nowhere* (Oxford: Oxford University Press, 1986).
25 Lamarque, P. and Olsen, S. H., *Truth, Fiction and Literature: A Philosophical Perspective* (Oxford: Oxford University Press, 1994).

26 Einstein, 'Reply to criticisms', p. 669.
27 Einstein, A., 'Letter to Max Born, 4 December 1926', in *The Born-Einstein Letters*, I. Born (ed.) (London: MacMillan, 1971) p. 9.
28 Sudbery, A., 'The Everett-Wheeler interpretation and the open future', in G. Jaeger, A. Khrennikov, M. Schlosshauer and G. Weihs (eds.), *Advances in Quantum Theory* (Melville, NY: AIP, 2011) pp. 492–97.
29 Sudbery, A., 'Philosophical lessons of entanglement', in *75 Years of Quantum Entanglement: Foundations and Information Theoretic Applications*, D. Home, G. Kar and A. S. Majumdar (eds.) (Melville, NY: AIP, 2011), pp. 7–14.
30 Aristotle, *De Interpretatione*, Chapter 9.
31 Einstein, A., 'Autobiographical notes', in *Albert Einstein: Philosopher-Scientist*, Schilpp (ed.), pp. 1–95, on p. 85.
32 Byrne, P., *The Many Worlds of Hugh Everett III* (Oxford: Oxford University Press, 2010), p. 221.
33 Ibid., p. 168.

5 Toward relational reality

From Einstein and Tagore to Gaudiya Vaishnava Vedanta

Ravi V. Gomatam

Introduction

Einstein and Tagore discussed, in two personal meetings (14 July and 19 August, 1930) the topics of Truth, Beauty and Reality.[1,2] Einstein began by posing the following question:

Einstein: There are two different conceptions about the nature of the universe: (1) The world as a unity dependent on humanity. (2) The world as a reality independent of the human factor.
Tagore: When our universe is in harmony with Man, the eternal, we know it as Truth, we feel it as beauty . . .
Einstein: This is the purely human conception of the universe.
Tagore: There can be no other conception. This world is a human world . . .
Einstein: Truth then, or Beauty is not independent of Man?
Tagore: No.
Einstein: I agree with regard to this conception of Beauty, but not with regard to truth.[3]

Taken at face value, the two disagreed over the nature of truth or reality. Einstein believed that truth is valid and reality exists, independent of human beings. Tagore, on the other hand, believed that the notions of truth and reality are entirely human. However, this discussion between two of the most outstanding minds of the last century deserves deeper study. There are already a considerable number of glosses on the dialogue between these two stalwarts (see, for example, Jammer[4] and Gosling[5]).

I shall not attempt to review all extant works, but proceed straight to my own, hopefully novel, gloss on their dialogue, from the viewpoint of issues in quantum mechanics (QM). Einstein's concern with realism and causality in quantum mechanics is well known. Tagore, too, was quite aware of the issues about realism and causality in quantum mechanics, as demonstrated in *Our Universe*, where he devotes a full chapter to the 'world of atoms' and discusses the evolution of quantum theory. Indeed,

Tagore raises the topic of causality and QM in his second dialogue with Einstein.

Tagore: I was discussing with Dr. Mendel today the new mathematical discoveries which tell us that in the realm of infinitesimal atoms chance has its play; the drama of existence is not absolutely predestined in character.

Einstein: The facts that make science tend toward this view do not say good-bye to causality.

Tagore: Maybe not, yet it appears that the idea of causality is not in the elements, but that some other force builds up with them an organized universe.

Einstein: One tries to understand in the higher plane how the order is. The order is there, where the big elements combine and guide existence, but in the minute elements this order is not perceptible.

Tagore: Thus duality is in the depths of existence, the contradiction of free impulse and the directive will which works upon it and evolves an orderly scheme of things.

Einstein: Modern physics would not say they are contradictory. Clouds look as one from a distance, but if you see them nearby, they show themselves as disorderly drops of water.

Tagore: I find a parallel in human psychology. Our passions and desires are unruly, but our character subdues these elements into a harmonious whole. Does something similar to this happen in the physical world? Are the elements rebellious, dynamic with individual impulse? And is there a principle in the physical world which dominates them and puts them into an orderly organization?

Einstein: Even the elements are not without statistical order; elements of radium will always maintain their specific order, now and ever onward, just as they have done all along. There is, then, a statistical order in the elements.[6]

Quantum mechanics can thus form a common ground for analysing their apparently divergent viewpoints to identify also a convergence. Tagore's viewpoint, as expressed in *The Religion of Man*, is that an individual can arrive at true knowledge of Reality only by perfecting his *relation* with the Universal Mind. Einstein's viewpoint was that science can help us grasp reality as it is, independent of our own existence, only that with QM, we need a drastically different conception of causality. I will argue that the ideas of Einstein and Tagore can be synthesised from the perspective of an Indian *Vedantic* tradition called the *Gaudiya Vaishnava Vedanta* (GVV), which features a *relational* conception of *matter*. To this end, I shall begin by analysing their respective positions: Tagore's notion of relational truth via his 'religion', and Einstein's demand for objective truth in science.

Tagore's view in *The Religion of Man*

In their first dialogue, toward the very end, Tagore refers to his 1931 book, *The Religion of Man*, as an elaboration of the thoughts he has expressed. The book was a result of the 'Hibbert Lectures' that he had delivered a few months prior to the two dialogues at Oxford in May 1930. In this book, Tagore explicitly states, 'Men of Science tell us that truth, unlike beauty and goodness, is independent of our consciousness. They explain to us how the belief that truth is independent of the human mind is a mystical belief, natural to man but at the same time inexplicable'.[7] This is the same argument that Einstein made in the dialogues. Thus, the points made in this book by Tagore are essential background for understanding and interpreting Tagore's statements in the dialogues.

However, we also need to take into account other writings of Tagore, such as in his book *Personality* to fully understand his views. In considering his views, Jammer writes, 'like Einstein, Tagore too did not find [his] religion in scriptures'.[8] Tagore himself states that his brand of religion has 'not grown through any process of philosophical reasoning [but] has followed the current of my temperament from early days until it suddenly flashed into my consciousness with a direct vision'.[9]

Nevertheless, in the first dialogue, Tagore refers to Indian philosophy:

> According to Indian Philosophy there is *Brahman*, the absolute truth, which cannot be conceived by the isolation of the individual mind or described by words but can only be realized by completely merging the individual in its infinity. But such a Truth cannot belong to Science. The nature of Truth which we are discussing is an appearance—that is to say, what appears to be true to the human mind and therefore human, and may be called Maya or illusion.[10]

Elsewhere, Tagore also refers to the *Upanishads*. Tagore's reference to 'Indian philosophy' in the dialogue certainly is scriptural, for *Brahman* is espoused in a central text of the Hindu tradition called *Vedanta* (a set of 570 aphorisms in Sanskrit stated by the sage, Krishna Dvaipayana Vyasa in BCE, well before the Hellenistic tradition). Gosling opines that Tagore refers to a specific interpretive version of *Vedanta*, known as the *Advaita* (non-dualistic) or monistic school of *Vedanta*. Gosling writes, 'one might conclude that when he is arguing with Einstein seen as a representative of Western science, Tagore counters what he perceives as western dualism with a more *Vedantic* insistence on unity'.[11] However, there are also non-monistic interpretive schools of *Vedanta*, known as *Vaishnava Vedanta*. Tagore allied himself more with the *Vaishnava* school.[12] Herein lies the context for understanding Tagore's part in the dialogues and trying to synthesise the positions of these two thinkers. The *Gaudiya*

Vaishnava Vedanta school that I have mentioned already dates from the 16th century and was propounded by Saint Chaitanya. The word '*Gaudiya*' refers to Bengal, the place where Tagore hailed from. Given that Tagore allied himself with the *Vaishnavas* in talking about the Supreme Person, it will be relevant to explore the resources of GVV to synthesise their two positions.

The Brahman of the *Advaita* school is pure spirit, the foundation of pure consciousness which is the generating seed of all phenomenal variety. It is described as spiritual light in the *Upanishads*. Tagore's vision of creation, however, echoes a more modern scientific cosmology:

> Light, as the radiant energy of creation, started the ring-dance of atoms in a diminutive sky, and also the dance of the stars in the vast, lonely theater of time and space. The planets came out of their bath of fire and basked in the sun for ages.[13]

The *Advaita* Vedantic vision of reality is non-theistic but spiritual: a unitary consciousness in which matter is ultimately unreal. In *The Religion of Man*, Tagore takes the reality of matter seriously. The *Vaishnava* school, while seeing *Brahman* or spirit as transcending matter, treats matter as objectively real. Thus, relating Tagore's position to the *Vaishnava* school makes possible a productive interaction with science.

Tagore also takes the scientific evolutionary view on board; in his religious worldview, Life progresses from a single cell with proto-consciousness to fully conscious man (the generic human):

> They [radiant energy, space, time, and the planets] were the thrones of the gigantic Inert, dumb and desolate, which knew not the meaning of its own blind destiny and majestically frowned upon a future when its monarchy would be menaced . . . Then came a time when life was brought into the arena in the tiniest little monocycle of a cell . . . it was made conscious, not of the volume but of the value of existence, which it ever tried to enhance and maintain in many-branched paths of creation . . . But the miracle of creation did not stop here in this isolated speck of life . . . a multitude of cells were bound together into larger unit, not through aggregation, but through a marvelous quality of *complex interrelationship* maintaining a perfect co-ordination of functions . . . Before the chapter ended Man appeared and turned the course of this evolution from an indefinite march of physical aggrandizement to a freedom of a more subtle perfection.[14] (italics added)

At the same time, Tagore goes past a purely evolutionary account of man's place in nature. The 'end of chapter' signified by the appearance of the human species is also the beginning of a phase beyond physical evolution, which talks of the universe as conceived from man's limited individual viewpoint:

> The processes of evolution, which after ages has reached man, must be realized in its unity with him; though in him it assumes a new value and proceeds in *a different path*. It is a continuous process that finds its meaning in Man; and we must acknowledge that the evolution which science talks of is that of Man's universe.[15] (italics added)

Tagore aims to go beyond the reductive, purely physical framework of modern science. He takes the stand that although man appears to be the culmination of purely physical evolution, his 'personality' calls for a unique task, that of achieving a greater unity *in relationship* with the universe. This role lies beyond the ken of Darwinian evolution and also beyond the ken of science.

Thus, Tagore's position is an admixture of scientific evolution of the species and a purely personal evolution in which each individual human has to perfect his/her unity with the larger whole: 'This spiritual process is not so easy as the physical one in the physical world'.[16] For Tagore, the physically embodied man, the result of chemical and Darwinian evolution, is mortal. But when he struggles and achieves unity with the rest of existence that is not him, he becomes immortal: 'His multicellular body is born and it dies; his multi-personal humanity is immortal'.[17]

Tagore also imbues the Universe with a Personality, referring to the 'Universal Man' or 'Universal Mind'. For Tagore, 'Truth [with a capital T] [is] the perfect comprehension of the Universal Mind.'[18] Each of us 'has an inner faculty of our own which helps us to find our relationship with the supreme self of man, *the universe of personality*'.[19] As we shall see later, he draws from the *Upanishads* to equate this personality with the 'Supreme Person' of the *Vaishnava* religion and to motivate the idea of man's search for his inner divinity as perfecting his relation with the Supreme Person. On the one hand, Tagore wrote of the materialistic universe that radiant energy gave rise to its planets in space and time, constituting the 'thrones of the gigantic Inert', which is 'dumb and desolate' and 'knew not the meaning of its own blind destiny'. On the other hand, he also wrote, 'it is not for man to be merely the recipient of favours from nature; he must fully radiate himself out in his creation of power and perfection of love. His movement must be towards the Supreme Person, whose movement is towards him'.[20] 'My religion is in the reconciliation of the Super-personal Man, the Universal human spirit, in my own individual being'.[21] In this quest, insofar as man remains an individual, his truth is limited. Insofar as he can comprehend Truth as it is, he has to cease his identity as an individual human being:

> If there be some Truth which has no sensuous or rational relation to human mind, it will ever remain as nothing so long as we remain human beings.[22]

Einstein is surprised by this view, and exclaims, 'Then I am more religious than you are!' Einstein's surprise is understandable, since he has written elsewhere:

> Everyone who is seriously involved in the pursuit of science becomes convinced that a spirit is manifest in the laws of the Universe—a spirit vastly superior to that of man, and one in the face of which we with our modest powers must feel humble. In this way the pursuit of science leads to a religious feeling of a special sort, which is indeed quite different from the religiosity of someone more naive.[23]

Einstein's experience as a physicist led him to believe in a transcendent ordering agency which he calls the spirit. This is Einstein's version of naturalised religion. Thus, he asks Tagore at the very start of their dialogue:

Einstein: Do you believe in the divine isolated from the world?
Tagore: Not isolated. The infinite personality of man comprehends the universe. There cannot be anything that cannot be subsumed by the human personality, and this proves that the truth of the universe is human.[24]

To explain the difference between man's limited individual truth and the Universal Man's Truth, Tagore gives an example:

> In the apprehension of Truth there is an eternal conflict between the universal human mind and the same mind confined in the individual. The perpetual process of reconciliation is being carried on in our science, philosophy, in our ethics. In any case, if there be any Truth absolutely unrelated to humanity then for us it is absolutely non-existing.
>
> It is not difficult to imagine a mind to which sequence of things happens not in space but only in time like the sequence of notes in music. For such a mind such conception of reality is akin to the musical reality in which Pythagorean geometry can have no meaning. There is the reality of paper, infinitely different from the reality of literature. For the kind of mind possessed by the moth which eats that paper literature is absolutely non-existent, yet for Man's mind literature has a greater value of Truth than the paper itself.[25]

Einstein, who clearly felt at this point of the dialogue that his view was very different from Tagore's regarding man's capacity to know Reality, five years later adopted a very similar musical analogy to express his thoughts about quantum causality:

> Our present rough way of applying the causal principle is quite superficial. We are like a child who judges a poem by the rhyme and knows

nothing of the rhythmic pattern. Or we are like a juvenile learner at the piano, just relating one note to that which immediately precedes or follows. To an extent this may be very well when one is dealing with very simple and primitive compositions; but it will not do for the interpretation of a Bach Fugue. Quantum physics has presented us with very complex processes and to meet them we must further enlarge and refine our concept of causality.[26]

Here Einstein emphasises that QM is compatible with *statistical* order, but he was personally more concerned with the fact that QM is not compatible with causality at the *single-event* level due to superposition. Since this feature of QM directly concerns realism as Einstein discussed it with Tagore, it is very relevant to turn our attention to Einstein's objections to quantum mechanics.

Einstein and quantum mechanics

To inquire about the nature of the physical reality that QM portrays is to face profound conundrums. As late as 1935, a full decade after Schrödinger's wave mechanics were established, Max Planck, the originator of the quantum postulate, wrote, 'There is no doubt whatsoever that the stage at which theoretical physics has now arrived is beyond the average human faculties, even beyond the faculties of the great discoverers themselves'.[27] Recently, Nobel Laureate Steven Weinberg remarked that a more satisfactory explanation of QM is still needed.

Einstein, too, had reservations about quantum theory from the beginning, even though he is one of its principal founders. He continually refined his concerns over the unsatisfactory aspects of QM, or at any rate, our understanding of it. Einstein acknowledged the tremendous practical successes that QM had brought: All of modern chemistry and electronics, for example, is based on QM.[28] Despite the stunning practical successes of QM, Einstein always held that 'an inner voice tells me it is not the real thing'.[29] He accepted it as a correct theory at the level of ensembles, but he believed a new theory was needed to get at the real situation pertaining to individual particles and systems.

This is not the place to go into a full review of Einstein's objections to QM. I have done it elsewhere.[30] In this chapter, I will be discussing only those aspects of Einstein's position relevant to the Tagore-Einstein dialogue.

The central point we need to consider is superposition, which poses severe problems to causality and realism, conceived along conventional lines. If ψ_1 and ψ_2 are two possible solutions to the Schrödinger equation, then any linear combination of these states ($a_1\psi_1 + a_2\psi_2$, where the coefficients are complex numbers) will *also* be a solution. Such solutions are known as states of superposition. Such a superposed state yields, under Born's rule, the probabilities for ψ_1 and ψ_2 to be observed at any given time. ψ_1 or ψ_2 is considered to be the state of the system *at* the point of measurement. But the superposed state itself does not correspond to any actual observation outcome. For this reason, Einstein saw quantum theory as 'incomplete':

> One would very much like to say the following: Ψ stands in a one-to-one correspondence with the real state of the real system . . . If this works, I talk about a complete description of reality by the theory. However, if such an interpretation doesn't work out, then I call the theoretical description 'incomplete'.[31]

Classically, given a probabilistic description, one could assume that nature determines a single outcome, only that our knowledge is incomplete. But, in QM, superposition is postulated to be a complete representation of the state of the single system. Yet, the superposed state itself cannot causally explain why, in a single measurement, the outcome is what it is. For this reason, Einstein saw quantum theory as descriptively 'incomplete' at the level of the single system, although he conceded that QM was predictively complete at the ensemble level:

> This theory is until now the only one which unites the corpuscular and undulatory dual character of matter in a logically satisfactory fashion; and the (testable) relations, which are contained in it, are, within the natural limits fixed by the indeterminacy-relation, *complete.* [italics in the original][32]

Einstein's concern was causality at the single-system level. He rejected the classical hidden-variables programme as a way of recovering causality within quantum theory. He wrote, 'I think it is not possible to get rid of the statistical character of the present quantum theory by merely adding something to the latter, without changing the fundamental concepts of the whole structure'.[33] Accordingly, Einstein expected a future, more 'complete' quantum theory to be fundamentally different from the current theory.

It is also necessary to stress that Einstein was not dissatisfied with QM just because of probabilities. In conjunction with superposition, quantum probabilities entail that single events are random and so cannot have a cause. Given this situation, the popular view is that Einstein said, 'God does not play dice' and thus could not accept the probabilistic quantum mechanics. Even Weinberg repeated this in his 2005 article in *Physics Today*, on the occasion of the centennial of Einstein's *Annus Mirabilis*. In response, I had pointed out that the full quote from Einstein is actually:

> That the Lord should play with dice, all right; but that He should gamble according to definite rules, that is beyond me.[34]

In other words, Einstein had no problem with individual events being truly random in nature. In such a case, he only expected that that an ensemble of events should also be random. In for the penny, in for the pound. But when an ensemble of events forms a regularity such as an interference pattern, Einstein argued that there must also be causality at the individual level.

Einstein's stance is appealing. If someone is throwing inked dots on the wall randomly, we do not expect a large number of throws to leave a well-drawn pattern, say, a picture of Abe Lincoln, do we? If that happens, we would have to suspect that there must be a hidden skill of the person involved, which would not be captured by remaining within the idea that he is mechanically 'throwing' the ink on the wall. We would have to think in terms of the person effectively 'painting'.

Despite embracing randomness at the level of single events, present QM succeeds spectacularly at predicting the behaviour of ensembles. But Einstein believed that we can go farther. As quoted earlier [25], he saw the quantum statistical order as being suggestive of a deeper conception of causality operating behind quantum phenomena.

He alludes to the same idea in his conversation with Tagore: 'Clouds look as one from a distance, but if you see them nearby, they show themselves as disorderly drops of water'.[35]

The belief in single-event causality, not provided by quantum theory, but hinted at by the emergence of non-random patterns out of single events, stopped Einstein from embracing Tagore's view that all human ideas about reality must remain subjective.

There is another good reason why Einstein thought that we must go beyond present QM to get the underlying causality. If we assume that superposition states in QM directly correspond to real states in the world, then we arrive at a conclusion even more counter-intuitive than the randomness of single events: that quantum reality *changes* at the point of measurement. The situation can be roughly summarised as follows.

Pre-quantum or classical physics is compatible with the assumption that the value we get when we perform a measurement on a system corresponding to a particular property such as position was possessed by the system *prior to measurement*. A classical measurement can be taken to passively reveal a pre-existing value. In QM, however, there is no such thing as a *passive* measurement. As Bohm astutely put it,

> [T]he properties of position and momentum are not only incompletely defined and opposing potentialities, but also that in a very accurate description, they cannot be regarded as belonging to the electron alone; for the realization of these potentialities depends just as much on the systems with which it interacts as on the electron itself.[36]

In QM, the process of measurement plays a fundamental role in the coming about of the observed state, in that the superposed state has to *change* into the observed (eigen)state at the point of measurement. Yet, no mechanism for this change in the wave function is offered or explained by QM, or even required by QM, to make its predictions.

Under the circumstances, the superposed state (which involves complex amplitudes) entails a *shift* from the language of real objects to the language of possibilities. Quantum 'objects', if we can even call them objects, have a peculiar nature. The objects of quantum theory, represented by wave functions, when presented with the possibility of going through one of two paths, as in the two-slit experiment, *seem to go through both paths simultaneously.* This causes extreme violence to our everyday intuitions.

However, the problem is deeper. The photon is *not* 'in' both paths, traversing them simultaneously. Rather, since all possible locations of the photon on the two possible paths are represented by complex numbers, the best explanation of the photon's state is that somehow, the *possibilities* for these locations have a reality, not the locations themselves. The locations are only realised when the experiment requires it. The quantum 'possibilities' are a state just as real as the localised observed state, but in some sense, that we do not understand yet.

Thus, it is not enough to say, as Tagore did, that a 'table' conceived as a solid object is only an appearance. Pre-quantum or classical physics does propose that the ordinary objects of our perception are really clusters of atoms vibrating about their mean positions. But what about those atoms? Are they real? Quantum reality says they are also 'appearances', in the sense that we can treat them as objects in the everyday sense only at the point of a measurement, and that too in a limited manner. Quantum reality is not just another level of nested classical reality giving rise to everyday appearances. It is a fundamentally different kind of reality wherein physical states refer to *possibilities* rather than objects.

There is a final sting at the end of the tail. QM does *not* limit these strange superposition states of possibility to the atomic realm. QM applies to the macroscopic realm also: In principle, *macroscopic objects can exist in superposition states*! The usual response is that in practice, macroscopic objects behave classically due to the large value of the mass and small value of the de Broglie wavelength. But, *in principle,* the Schrödinger evolution is applicable to the all regimes, including the macroscopic, and that equation does not require the superposition to ever go away. This is why Einstein poignantly asked, 'Does the moon exist only when we observe?'[37] Indeed, examining the consequences of *extending* current QM toward the macroscopic regime to test the limits of superposition is now an active area of both theoretical and experimental research.

QM, in the hands of physicists, is much like how most people drive a car without knowing how the car works or watch TV or use the cell phone without knowing how these devices work. Nobel Laureate Max Born remarked, 'At present, the theory is more intelligent than us. It knows why it works, we don't'. Decades later, in the 80s, another Nobel-prize winning quantum physicist, Murray Gell-Mann, was exasperated: 'Quantum theory, that strange thing we all know how to use, but understand very little'.[38]

How have quantum physicists managed to put the theory to such tremendous practical use without 'understanding' how it works? We should remember that the issue is 'understanding' how a *single* quantum system works. However, the behaviour of an ensemble of quantum systems is perfectly amenable to statistical treatment. Thus, the so-called 'standard interpretation' of QM is silent about the real states of single systems. At best, there is the idea of a *mathematical representation* of the *quantum state* of the single system that has no one-to-one correspondence with the 'real' state of the single system.

To justify avoiding the issue of the real state of single particles, the standard interpretation introduces a further idea that QM assigns a definite value (called an eigenvalue) to a property only *at* the point of a measurement, and that within QM, there is no room to talk about values in a measurement-independent context. One has to know a lot of details about QM in order to see how all these ideas combine together to form a pragmatically successful scheme of application without worrying about all the single-system paradoxes we spoke about earlier.

But this compromise is the very thing that Einstein abhorred. He called it 'an attitude fit only for engineers and shopkeepers' (without intending, I am sure, to put down these two categories). His belief was that making successful predictions constituted certainly half, but only half, of the task of science. It is 'understanding' the reality that is the real aim of physics.

We can now see why Einstein felt he could not agree with Tagore's proposal that humanity, singly or collectively, can reach only a subjective mental conception of truth, not truth as it is.

Einstein: If there would be no human beings any more, the Apollo of Belvedere would no longer be beautiful.
Tagore: No.
Einstein: I agree with regard to this conception of Beauty, but not with regard to Truth.
Tagore: Why not? Truth is realized through man.
Einstein: I cannot prove that my conception is right, but *that is my religion.*[39]

Tagore's idea of truth being human is too close to the operational viewpoint of the standard interpretation of QM that Einstein fought all his life: 'What I dislike in this kind of argumentation is the basic positivistic attitude, which from my point of view is untenable. It comes to the same thing as Berkeley's *esse est percipi* [to exist is to be perceived]'.[40]

However, Tagore did not see himself as denying the reality of worldly objects if he himself were to die: 'If a particular man as an individual did not exist, the table would exist all the same, but still as a thing that is related to the human mind'.[41] Tagore sees the 'table' as a human construct, perhaps only in the sense that it is a notion of reality that functions only when we admit a dualism between our individual selves and the larger unity called Reality. Once man achieves his full divinity, such a conception of reality (in terms of individual objects) would evaporate. How does the Superman view reality, if not in terms of objects in the external world? Tagore does not elaborate on this in the dialogue. But that would be of great relevance to QM, which denies the very reality of the external-world-object concept.

If neither Einstein nor Tagore denies human reality (with a small 'r'), and if both believe in the objective existence of reality independent of individual minds, the only issue is how far Tagore's ideas can be relevant to quantum mechanics, the main topic of interest to Einstein. The common point between their two belief systems is the part/whole relationship between individual man and Universal Man, or between individual bright spots on the photographic plate and the 'interference pattern' they combine to form.

I now turn my attention to developing this idea further.

Toward a synthesis

Both Tagore and Einstein agreed that there *is* a larger Reality, of which individual persons are only a part. They also agreed that humans can and must know the truth of this Reality. Tagore took the position that individual humans are limited to an imperfect, subjective conception of Reality as long as they have not perfected their *personal* relationship with the totality of Reality. He imbues Reality with a personality of its own that he calls variously Super Man, Super Mind or the Supreme Person. An individual person can know Truth only by perfecting his relationship with the Superman, at which point he no longer maintains his own separated, individual perspective. That was Tagore's 'religion'.

Einstein believed that Reality exists independently of our acts of observation, i.e., independent of any personal relationship with it, and that it can be known as such by individual humans. He believed that science is the way to attain this Truth, and said, 'I cannot prove that my conception is right, but that is *my* religion'.[42] Elsewhere, Einstein called this his 'epistemological credo'.[43]

Einstein wrote, "I believe physics concerns the 'real', but I am not a realist".[44] Einstein had his own brand of realism.[45] I too am a realist, and I share Einstein's belief that science can aid in getting at a view of Reality that is independent of our consciousness, but I do not accept his idea that to be objective, the scientific conception of reality must be independent of measurements. For if we are part of reality and come to have a certain perspective on reality, how can that perspective *not* be part of reality? Similarly, if in QM, our measuring devices are as much a part of the physical reality as the systems upon which measurements are made, how can their interaction be *not* part of reality?

Indeed, I will argue that in quantum theory, a measurement can be objectively modelled, if we treat it as actualising an *irreducible relation* between the measured system and the measuring arrangement.

In this context, Tagore's emphasis on ultimate Truth being a 'relationship' between man and Reality becomes interesting. 'A multitude of cells were bound together, not through aggregation, but through a marvelous quality of complex *interrelationship* maintaining a perfect co-ordination of functions'.[46] This is very compatible with what quantum mechanics entails, namely that individual observation is not just related to a property of the observed system. The observing arrangement has an essential role in creating the observed value, as already explained. For this reason, in *opposition* to the Copenhagen Interpretation, Bohr emphasised a corresponding 'inseparability' between the measured system and the totality of the experimental arrangement. As we shall see, this opens the room for introducing a new, relational view of properties within QM.

Tagore on man's relationship with man

Relationship is the fundamental truth of this world of appearance.

—Tagore[47]

Tagore emphasises that the personality of man is more evolved compared to that of animals[48] and other forms of life. Tagore sees achieving the full potential of this personality as the search for divinity, the special prerogative of man. But this cannot be achieved without going beyond an exploitative relationship with the world of matter. Admitting a relation with a Supreme Person in love is needed. He gives the analogy of the child in the womb, which is 'narrowly self-contained in its loneliness'. After birth, the child does not live merely by fixing its roots in the world of matter. Because it has personality, its full realisation needs a *relationship* with its mother. 'The freedom of personality [of the child] has its fulfillment, not in itself, but in other personality [mother], and *this is love*'.[49]

In a similar way, citing *Iso-Upanishad,* a Vedic scripture, Tagore takes the position that the unity of consciousness is the loving relationship between the individual man and the Supreme Person.[50] This Supreme Person is 'the centre of all relationships' and is 'the ultimate object of attainment'.[51] He goes on to say that in India, 'the *Vaishnava*s have realized this truth and have boldly asserted it'.[52] However, he follows this up by saying, 'God has to rely on human souls for the fulfillment of his love'.[53] From my own training and perspective on the GVV tradition of India, I would put that the individual soul has to rely on God to fulfill its propensity to love, which is unproductively directed towards external world objects or towards negating one's very self by seeking monistic liberation of merging in the Supreme. GVV calls for the individual to redevelop an identity of the self that is entirely defined in its relation to the Supreme Person, not in relation to the body or world. This requires going past both the transcendental monistic *Vedanta* perspective of unitary consciousness without individuality and also the materialistic monism of modern science that limits man to his embodied individuality. Such a viewpoint is compatible with Tagore's emphasis on the development of the matter-transcending aspect of human personality, seeking its fulfillment in the relation of love with the Personality of the Universe (*Vishwatman*). However, unlike Tagore's system, GVV presents matter also in terms of such a relational ontology, which makes it potentially relevant to the scientific issues of causality in QM, which Einstein was interested in.

In classical mechanics, too, measurements represent a relation between the system under observation and the measuring apparatus, and the values obtained in a measurement quantify such relations. For example, to say that an object is 3 feet high captures a particular relation between the object and the particular scale we have used. All of classical physics embodies such a relational character of measurement using scales and clocks.

But such a relational description can be called epistemological, in the sense that both the measured object and the measuring device are believed to be real in the external world and relations between them are what we impose upon them. However, as we have already seen, QM requires treating quantum objects in terms of possibilities or potentialities to bring about certain measurement results. But the measuring devices themselves are taken to have a real existence in the external world. I will argue that if we limit even macroscopic objects to have *only* existence in appearance, i.e. as phenomenal objects, then we can treat relations (between space-time regions) as ontologically real and objects as being epistemological. This will be the *reverse* of the classical mode of relational description.

GVV presents such an ontological notion of matter, and this is particularly presented in its component known as *Bhaghavata Sankhya*, which is quite different from the *Iswara Krishna Sankhya* more commonly discussed amongst Indologists. The *Bhaghavata Sankhya* goes into considerable detail about an objective, relational ontology of matter. I shall now summarise some relevant aspects of the GVV conception of matter. Bearing in mind space limitations, I shall not cite any source material.

The Gaudiya Vaishnava Vedanta tradition and a relational ontology for matter

GVV is a dualistic framework. However, the labels 'dualistic' and 'monistic' are used by Indian philosophical traditions in several different contexts. Both dualistic and monistic traditions accept that the ultimate reality is unitary *Brahman*, pure consciousness. However, while the monistic *Advaita* tradition holds that *Brahman* is only impersonal, GVV holds that *Brahman* is *both* personal and impersonal. This is one reason it is called a dualistic *Vedanta*. GVV describes the personal aspect of *Brahman* in the most exhaustive manner. It is this personal conception of *Brahman* that Tagore embraced, following the Baul tradition. The second aspect of GVV's dualism is over the question whether our individual existence is illusory or real. *Advaita* holds that individuals are temporary identities, ultimately to be transcended and given up. The dualist tradition says that both the Supreme *Brahman* and the individual *Brahman* are eternal individuals, both transcending matter.

The third dualism addresses the important question: Does 'matter' exist apart from *Brahman* as a separate ontological category? The monists deny this: They say the absolute plane is unitary and only *Brahman* exists. The GVV holds that *Brahman* as a Person is unitary and represents the energetic aspect, while *Brahman* as energy has two broad categories: *chit* and *achit*, consciousness and matter. Without going into details, what is important is that because GVV *accepts matter as objectively real*, this tradition is potentially more relevant to science than the *Advaita* tradition. In GVV, knowledge of matter is an essential part of knowing the ultimate Truth.

The distinction in the doctrine of *Maya* between the monistic *Vedanta* and GVV is also very pertinent. In monistic *Vedanta*, the material world is illusory and matter itself is illusory. What exists is only Spirit, or Consciousness. In GVV, there *is real* matter, and only the way we perceive it is illusory. That is to say, *matter appears different from what it is*. This realistic view of matter is very compatible with quantum theory.

But GVV proposes a unique, even more outstanding idea about matter that is particularly relevant to the Tagore-Einstein debate. As already mentioned, our current common-sense notion is that objects are ontological and relations are epistemic, whereas GVV *inverts* this conception. At the level of spirit, although the Supreme Person is the original basis of all existence, *Bhakti*, or the *relation* between the individual and the Supreme Person, is ontologically *higher* than *both*, though all three are ontologically real. GVV offers a parallel to this theological idea in its conception of matter: The relational properties of matter are ontological, while the objects themselves are given only in perception. Thus, the ontology of real macroscopic objects in the external world is renounced. Hence, the world of matter is only appearance, *maya* (ma = not; ya = what it appears to be). In GVV, matter is real, but the material world is an appearance.

Relational properties for matter

The essence of a relationship is that its relata contribute *equally*. In QM, eigenvalues (the observable values of a measured physical property) can be seen as representing the magnitude of a *relation* between the measured system and the measuring arrangement because the set of available eigenvalues in any given experiment depends equally on both. This idea is different from but complementary to the textbook interpretation of contextuality in QM, which states that a measured property belongs to the observed system, but only in the context of a well-defined measurement. Bohr was unique among the founders of QM in interpreting the quantum measurement as relational:

> The quantum postulate implies that any observation of atomic phenomena will involve an interaction with the agency of observation not to be neglected. Accordingly, an independent reality in the ordinary physical sense can neither be ascribed to the phenomena nor to the agencies of observation.[54]
>
> No result of an experiment can be interpreted as giving information about independent properties of the objects, but is inherently connected with a definite situation in the description of which the measuring instruments interacting with the objects also enter essentially.[55]

As quoted earlier, Bohm brings up the same point. But as both Feyerabend[56] and Jammer[57] pointed out, Bohr did not shift his practice of quantum mechanics to focus on relational properties.

Elsewhere, I have discussed the difference between Bohr's interpretation and the standard interpretation, often called the Copenhagen Interpretation,[58] and the need to introduce relational properties to interpret quantum mechanics.[59] A relevant excerpt is the following:

> Let us consider a macroscopic object, say a book. From the viewpoint of classical mechanics, such an object would be described by a set of numbers representing mass, position, velocity, etc. Since classical theory is deterministic, these numbers are presumed to exhaust all causally relevant properties of the object. Characterizations of the object as a book, or as a 'gift from my father' are taken to be epiphenomenal descriptions, extraneous to physics.
>
> However, in a consistent interpretation of quantum theory, if the macroscopic world would also have a quantum description, then a macroscopic object must possess more physical properties than those accounted for by classical mechanics. Again, an empathetic consideration of ordinary experience suggests some starting clues. A book can be used as a paperweight or a doorstop. Normally, these would be regarded as different uses of an object. However, they can be considered as involving a new kind of objective property, namely a relational property.
>
> The use of the book as a paperweight requires setting it up in a particular spatio-temporal relation with another physical object, i.e. placing the book on top of a stack of papers. Thus 'paperweight-ness' can be regarded as a potential property of the object, which becomes 'physically real' only when the object is placed in an appropriate spatio-temporal relation with another object. Yet, the property itself objectively belongs to the book. Let us call such a property a 'relational property'. It is similar to and yet different from primary properties that

> physics has so far studied. Both primary and relational properties are objective in the sense that both are defined in relation to another object, and thus can be said to exist in the object independent of the existence of conscious observers. However, as the ensuing discussion will try to show, the two are different in the sense that while the primary properties are only expressed by a relation with another object (such as a scale or a clock), the relational properties are expressed by and actualized in a relation with another.

The above discussion needs to be supplemented in one important respect. The idea of relational properties has been motivated above by giving examples of relations between *objects.*

Indeed, current QM retains the idea that macroscopic objects—measuring instruments—are real in the external world and describes the behaviour of quantum objects in the context of the behaviour of measuring devices. This approach has yielded immense pragmatic success, but is ontologically infertile since (as I am espousing here) understanding quantum reality may require renouncing the reality of even macroscopic objects in the external world. I believe that physics needs to shift to a complementary view of quantum theory wherein the relations between space-time regions are ontologically real and objects *even at the macroscopic level are* only epistemologically given, not real in the external world. This is the idea of relational reality I am moving towards. It is beyond the scope of this chapter to elaborate on this research programme, which is currently under development.

The GVV concept of *elements*, which is quite different from that of the Greeks, talks of material objects arising out of relational percepts. A preliminary account of the *Gaudiya Vaishnava* Ontology of God, soul and matter, without going into all the theological details, can be found in a monograph by the physicist Henry Stapp.[60] A more detailed account of GVV in full can be found in the work of Prabhupada.[61] A forthcoming work of mine is also relevant in this regard.[62]

Conclusion

Einstein brought an important conviction to the dialogues: Through science, we can give an 'objective' description of reality, as it is thought to exist independent of us. However, he conflated this objectivity that is independent of subjects with the demand that physical theory should be independent of a causal role for measurement. Measurement, being an objective interaction involving the measured system and measuring agency, need not undermine objectivity if the relational character of measurement is brought to the fore within QM in an appropriately new way. Einstein's conception of objectivity precluded this option.

Tagore took the opposite position that science, like every other human endeavour, will always give a limited view of Reality or Truth. However, he brought another strong conviction to the dialogues: that Truth can only be relationally comprehended. He presented this idea from the viewpoint of his religion, which he defined as the effort to perfect one's relation with the Supreme Person through devotion to transcend one's limited individuality. In this, Tagore allied himself with the Indian Vaishnava tradition that focuses on the devotional relation between man and Supreme Person. But he did not bring his notion of relational conception of truth to bear upon the concept of matter, which is at issue in QM, whereas the *Gaudiya Vaishnava Vedanta* tradition of Bengal, Tagore's birthplace, extends the relational conception to matter too. GVV thus can potentially synthesise the respective positions of Einstein and Tagore. I have suggested that GVV can therefore open the room for recovering objectivity and realism in quantum theory by the idea of a relational reality, if we are willing to move from the classical idea of macroscopic objects as having existence in the external world to the quantum idea that they too exist only as a result of measurement. This way of interpreting QM is an ongoing project of mine that I have further elaborated in a number of other papers.[63]

Acknowledgements: I thank Greg Anderson, C.S. Unnikrishnan and Kanwaljeet Kaur Saluja for their useful comments on the penultimate draft.

Notes

1 Henceforth, the 'dialogues'.
2 Tagore, R. and Einstein, A., 'Note on the nature of reality', in *The Religion of Man*, R. Tagore (ed.) Appendix II (London: George Allen and Unwin Paperbacks, 1931) pp. 221–25; see also Appendix 1 & 2 in this volume; Tagore, R. and Einstein, A., 'Three conversations: Tagore talks with Einstein, with Rolland, and Wells', *ASIA* 3 (1931), pp. 139–43, 196.
3 Tagore and Einstein, 'Note on the nature of reality', pp. 221–22.
4 Jammer, M., *Einstein and Religion: Physics and Theology* (Princeton, NJ: Princeton University Press, 1999).
5 Gosling, D. L., *Science and the Indian Tradition: When Einstein Met Tagore* (London: New York: Routledge, 2007).
6 Tagore and Einstein, 'Three conversations'.
7 Tagore, *The Religion of Man*, p. 12.
8 Jammer, *Einstein and Religion*, p. 71.
9 Tagore, R., *Personality* (London: Macmillan, Indian Edition, 1923), p. 11.
10 Tagore and Einstein, 'Note on the nature of reality', p. 223.
11 Gosling, *Science and the Indian Tradition,* p. 30.
12 'The *Vaishnava*s have realized this truth and have boldly asserted it'. Tagore, *Personality*, p. 102. Tagore also embraced many ideas of the Baul, who are not without connection to Vaishnavism, and Saint Chaitanya, who propounded the GVV.
13 Tagore, *The Religion of Man,* p. 1.
14 Ibid.
15 Ibid.
16 Ibid., p. 14.
17 Ibid., p. 10.
18 Tagore and Einstein, 'Note on the nature of reality', pp. 222–23.

19 Tagore, *The Religion of Man*, p. 11, emphasis added.
20 Tagore, *Personality*, p. 101.
21 Tagore and Einstein, 'Note on the nature of reality', p. 225.
22 Ibid., p. 225.
23 Dukas, H. and Hoffman, B., *Albert Einstein—The Human Side* (Princeton, NJ: Princeton University Press, 1979), p. 33.
24 Tagore and Einstein, 'Note on the nature of reality', p. 221.
25 Ibid., p. 225.
26 'Epilogue: A socratic dialogue between Planck, Einstein and Murphy', in *Where Is Science Going?* M. Planck (ed.) (New York: W. W. Norton and Co., 1932), pp. 203–4.
27 Ibid., p. 220.
28 The entanglement between two space-like separated particles invoked in his famous EPR paper has become the cornerstone for modern-day quantum teleportation experiments.
29 Born, M., *The Born-Einstein Letters* (London: McMillan, 1971), p. 71.
30 Gomatam, R., *Integrating Bohr and Einstein*, Ph.D. Dissertation, Department of Philosophy, Bombay University, India, 1998.
31 Letter to Schrödinger, 19 June 1935; cited in Fine, A., *The Shaky Game: Einstein, Realism and Quantum Theory* (Chicago: The University of Chicago Press, 1986), p. 71
32 Einstein, A., 'Reply to criticisms' in *Albert Einstein: Philosopher-Scientist*, P. A. Schilpp (ed.), 3rd edn. (La Salle, IL: Open Court, 1949, 1951), pp. 666–67.
33 Fine, A., *The Shaky Game*, p. 57
34 Einstein, A. quoted in Wheeler, J. and Zurek, W., *Quantum Theory and Measurement* (Princeton, NJ: Princeton University Press, 1983), p. 8.
35 Tagore and Einstein, 'Three conversations'.
36 Bohm, D., *Quantum Theory* (New York: Prentice Hall, 1951), p. 620.
37 Jammer, M., *The Philosophy of Quantum Mechanics* (New York: Wiley, 1974) p. 151.
38 Gell-Mann, M., What ARE the building blocks of matter?' in *The Nature of the Physical Universe*, D. Huff (ed.) (New York: Wiley and Sons, 1979).
39 Tagore and Einstein, 'Note on the nature of reality', p. 222.
40 Einstein, 'Reply to criticisms', p. 669
41 Tagore, *Religion of Man*, p. 13.
42 Tagore and Einstein, 'Note on the nature of reality', p. 222.
43 Einstein, 'Autobiographical notes', p. 13.
44 Einstein Archives, item No. 22–307, cited in Howard, D., 'Was Einstein really a realist?', *Perspectives on Science* 1 (July 1993), p. 225.
45 Gomatam, R., 'Einstein's realism', unpublished manuscript.
46 Tagore, *Religion of Man*, p. 1.
47 Ibid., p. 14.
48 Tagore, *Personality*, p. 98.
49 Ibid., p. 96.
50 Ibid., p. 99.
51 Ibid., p. 98.
52 Ibid., p. 102.
53 Ibid., p. 102.
54 Bohr, N., *Atomic Theory and Description of Nature* (Cambridge: Cambridge University Press, 1934), p. 54.
55 Bohr, N., *Atomic Physics and Human Knowledge* (New York: John Wiley and Sons, 1957), pp. 25–6.
56 Feyerabend, P. K., 'Niels Bohr's interpretation of the quantum theory', *Current Issues in the Philosophy of Science; Symposia of Scientists and Philosophers*,

H. Feigl and G. Maxwell (eds.) Proceedings of Section L of the American Association for the Advancement of Science, 1959 (New York: Holt, Rinehart and Winston, 1961), p. 372.

57 Jammer, *The Philosophy of Quantum Mechanics*, pp. 197–98.

58 Gomatam, R., 'Niels Bohr's interpretation and the Copenhagen interpretation—Are the two incompatible?' *Philosophy of Science* 74(5) (December 2007), pp. 736–48.

59 Gomatam, R., 'Quantum theory and observation problem', *Journal of Consciousness Studies* 6(11–12) (1999), pp. 173–90.

60 Stapp, H. P., *A Report on the Gaudiya Vaishnava Vedanta Form of Vedic Ontology* (Berkeley: Bhaktivedanta Institute, 1994).

61 Prabhupada, A. C. B. S., 'Fundamental principles of material nature', in *Srimad-Bhagavatam*, A. C. B. S. Prabhupada (ed.) Third Canto, Part 4, Chapter 26 (Los Angeles: Bhaktivedanta Book Trust, 1974), pp. 67–145.

62 Gomatam, R., *Matter and Consciousness in Bhaghavata Sankhya—A Contemporary Scientific Introduction*, forthcoming.

63 http://www.insist.ac.in/rgomatam

6 Science, poiesis and visions of reality

Tushar K. Sarkar

True logic is meta-logic, . . . True mathematics is meta-mathematics . . . likewise True physics is metaphysics.

C. F. von Weizsäcker

The recourse to metaphysics is like throwing a match stick into the powder magazine. It blows up the whole arena.

A. N. Whitehead

A creative scientist who aspires to get a good glimpse of reality needs to be a visionary, able to write poems about the cosmos.

Unknown

Unpacking the question

Every question anticipates and often evokes an answer. However, not all of them are equally relevant or significant answers—some fail to be either. Most often, it is due to the fact that the question being asked is unclear or imprecise or unfocused. It happens especially when the questions being asked (like, 'What is Reality?') are metaphysical in nature and cannot be given definitive answers strictly from within the confines of the empirical sciences. But this does not mean that the results obtained in different areas of empirical sciences are of no relevance to arriving at an answer. As a matter of fact, in this chapter, we will try to find an answer to our question viz., 'What is Reality?' on the basis of what can be gleaned from our present state of knowledge of 'physics' pertaining to its different fields of study. (I shall use 'physics' in its original broad sense of φ υ σ ι ς' to mean a scientific theory about any given aspect of the universe. In this broad sense, 'physics' would *not* exclude 'biology', 'cognitive sciences' etc. The context would make it clear if and when 'physics' is used here in a more restricted sense, e.g., in expressions like 'classical/Newtonian physics', 'modern physics', 'elementary particle physics' etc.)

Assuming that initially, we all agree about what 'Reality' means, the question inevitably prompts one to ask: (i) *How* does the nature of Reality (whatever it may be) come to be known? This brings the theory of knowledge or 'epistemology' into the picture.[1] (ii) Again, one may ask: Is *what* we know about the *nature* and *existence* of Reality (granted that we are somehow able to *know* it) independent of *how* we know it or is it not? This is an ontological question that *unavoidably* enters the scenario. Finally, (iii) there is the problem of the *interpretation* of a scientific theory. Even if we assume that there is a universal consensus about the respective answers to (i) and (ii) above and that everyone agrees that via a scientific theory, one can glean the *true* nature of Reality, we must still remember that a scientific theory is *not* like a faithful mirroring device of Reality. Any scientific theory, expressed as it must be in some appropriate language, needs to be *interpreted*, and it can always be interpreted in more ways than one. For example, the Newtonian concept of 'force' was used by physicists for centuries without there being any agreement about whether it is (or is not) something 'real'—whether or not 'force' is an *actually existing* furniture, so to say, of the external world. Du Bois-Reymond raised this issue in 1852, and according to Karl Pearson, '. . . *force* is a ghost of the old spiritualism'.[2] James Jeans says, 'Relativity has now shown that electric and magnetic forces are not real at all. Newtonian force of gravitation, energy, momentum all prove to be mere mental constructions, and do not even pass the test of objectivity'.[3] Again, Weizsäcker makes a subtle distinction between 'not exist at all' and 'not actually exist' with a view to meet some objections raised against quantum mechanics in the Einstein, Podolsky and Rosen (EPR) paper.[4] The examples cited above, irrespective of their merits, show one thing clearly viz., that even the most basic words like 'real', 'exist' etc., even when used in science, can be and have been interpreted in different ways. Therefore, interpretations play a crucial role in understanding scientific theories and subsequently, in making models of Reality on the basis of such interpretations. Again, it is possible that two pictures/models of Reality, as suggested by two different interpretations of a given theory, though equivalent from an abstract theoretical point of view, may still happen to be mutually incompatible. The Bohr-Einstein controversy may be viewed as a disagreement about the proper interpretation of the standard formalism of quantum mechanics (QM). Moreover, it may also happen that a given theory may be compatible with different, but mutually incompatible, micro-level interpretations. What happens then? Which picture, if any, of Reality are we then supposed to accept as *the* correct one? Leaving the fancy philosophical terms aside, all the three issues mentioned above (viz., those pertaining to epistemology, ontology and interpretation) are often rolled into one by the scientists and the philosophers of science under the heading 'methodology of science'. It follows that a new scientific theory may happen to be so much out of the ordinary that it forces some radical changes in the existing standard methodology of science. According to the Copenhagen school, such a radical change is required by QM. For example, if a methodology requires that for any description to be *objective*, 'it is necessary in almost every field of knowledge to pay attention to the circumstances under which the relevant evidence is obtained', then such a methodology not only differs from the pre-QM methodology, but it also changes the very meaning of 'objectivity' itself. If so, then such a radical change in methodology, in its turn, would drastically change the way we need to look at the questions about the nature of Reality itself. If and when such a change actually occurs, as it happens when the methodological changes suggested/implied by QM are followed, it is called a 'paradigm shift'. 'Methodology of science' and 'paradigm shifts' both evolve and develop via a sort of mutually interactive feedback loop. A paper shows, in detail, how intricately related an answer to the question, 'What is Reality?' is with methodological and interpretational presuppositions.[5] The examples of interpretational disagreements cited above are cases of what we propose to call '*theory-generated* problems of interpretation'. In contrast to these, there is another class of interpretational problems that belongs to a different level, in the sense that they originate at a level away/up from the level of *purely scientific* theories. Such interpretational controversies we call 'meta-theoretic problems of interpretation'. For example, EPR proposed the criterion of Reality: 'If, without in any way disturbing a system, we can predict with certainty the value of a physical quantity, then there exists an element of physical reality corresponding to this physical quantity' *vis-à-vis* claims like 'There is no quantum world' (Bohr), 'Atoms and elementary particles are not real; they form a world of potentialities . . . rather than one of things or facts' (Heisenberg) etc. This illustrates what I mean by 'meta-theoretic problems of interpretation'. Bohr said, 'There is only an abstract quantum mechanical description. It is wrong to say that the task of physics is to find out how Nature ***is***. Physics concerns what we can ***say*** about Nature'. We may or may not agree about it, but what we cannot disagree about is that Bohr's claim is meta-theoretic/meta-scientific in nature. Even a cursory look at the way the great minds formulate their respective positions indicate that the issues *cannot be* resolved (either in favour of, or against, any one of the views) from *within* the theoretical framework of science because, although both Einstein and Bohr agree about the 'correctness' of QM, they disagree about its 'completeness'. This difference of opinion ultimately boils down to their disagreement about the *aim/objective* of physics. Implicit in this debate is another profound meta-theoretic disagreement about the nature of 'scientific explanation'. As a consequence, while Einstein's concern is about the nature of 'physical Reality' and whether it really 'is/exists', Bohr's concern is exclusively about 'Nature-as-experienced' and 'what can be *said*' about it. We can safely presume that Bohr, due to his 'operationalist' leaning, somehow felt motivated by Wittgenstein's view, viz., 'The limits of my language are the limits of *my* world'.[6]

The problem in perspective

The Reality question gained prominence *within* the professional scientific community mainly after the advent of QM and its alleged 'incompleteness', supposedly *proved* by the 1935 EPR paper,[7] followed by a powerful rejoinder by Bohr himself.[8] Later, others joined the issue gradually. The controversy about the nature of Reality revolves around the following points:

(α) In what sense can there be an objective Reality?
(β) Is the EPR criterion of objective Reality acceptable?
(γ) Does the EPR objection strictly satisfy its own criterion of Reality?
(δ) Assuming that the EPR objection is fatal to QM, in what exact sense does it show the incompleteness of QM? In other words, how metaphysically loaded is Einstein's notion of incompleteness itself?
(ϵ) How basic and indispensable is Einstein's locality requirement?
(ζ) Is the measuring instrument to be considered a part of QM itself or, not?
(η) Can successful prediction be the sole requirement for *adequacy* of scientific explanation?
(θ) Can there conceivably be any experimental/formal way for deciding in favour of, or against, any one of the possible alternative answers to any of the above questions?

My endeavour here will be to put forward a *reasonably comprehensive* and adaptively flexible view about the nature of Reality while keeping the discussion, as far as possible, within the parameters laid down in terms of (α)—(θ) above.

Regarding (α), the Copenhagen school maintains that the notion of 'objective reality' makes no sense, but Einstein believed in the possibility of giving a model of Reality that shall represent *events themselves* and not merely 'the probability of their occurrence'. EPR proposed the following criterion for objective reality: If, without in any way disturbing a system, we can predict with certainty the value of a physical quantity, then there exists an element of physical Reality corresponding to this physical quantity. Their choice of two words, '*predict*' (instead of 'measure') and '*corresponding to*' (instead of 'identical with'), is very significant here. When I say, 'I see a *crayon* there', there *is*, for sure, an object that corresponds to/is denoted by 'crayon', but (i) it is certainly *not* identical with the word 'crayon' and moreover, (ii) it leaves unspecified what a crayon could be like. To a Frenchman and to a speaker of English, it would mean two different things. This leaves the EPR criterion somewhat ambiguous—an objection that Bohr had raised against the EPR argument.

Regarding (β), (γ) and (δ), it is now well documented that Einstein was dissatisfied with the EPR paper, which was written by Podolsky, because it obscured the central point.[9] In his *Autobiographical Notes*[10] and elsewhere, he argued that quantum mechanics implies that two systems A and B that are spatially separated and non-interacting can be in a *non-separable* (or *entangled*) state ψ_{AB} in which neither has an independent wave function. Now, if one makes a certain measurement M_1 on A corresponding to some observable O_1, then according to QM, the systems will no longer be non-separable and B will have a wave function, say ψ_{B1}. If, on the other hand, one had done a measurement M_2 on A corresponding to an observable O_2, which is incompatible with O_1, then according to QM, B would have acquired a *different* wave function, say ψ_{B2}. Now, let us *assume* that the *factual state* of B cannot be affected in any way by what is done to A which is spatially separated from A and also does not interact with A in any way. This is the *Einstein criterion of reality + locality*. Then, one has to agree that there are only two options open: either QM is incompatible with *reality + locality*, or QM does not assign a unique wave function to all systems and therefore must be incomplete. This is a theorem, and the precursor to Bell's theorem[11] that puts it in a testable form. Hence, only an empirical test can tell which option is right. Recent experiments favour the first option.

Regarding (δ), it should be pointed out that there are at least four different senses (as distinct from *semantic incompleteness* of QM, which I discussed earlier) in which a scientific theory may be said to be *ontologically incomplete*. The point of Einstein's complaint was that QM fails to satisfy the requirement of descriptive ontology. For the advocates of QM, it is not a defect at all, because QM challenges the very need of postulating a pre-given and observation-independent fixed ontology of the quantum world. It should be clear by now that the whole thing boils down to a clash between two disparate meta-scientific commitments, neither of which can be justified from *within* a scientific framework.

As regards (ϵ), we should remember that Einstein's unshakable conviction about *reality + locality* was the main basis of his criticism of QM. However, in 1952, David Bohm developed his hidden variable theory (HVT) of quantum phenomena with a *realist* orientation that does not satisfy the locality criterion.[12] Bohm's 'quantum potential' has a non-local aspect in the sense that it acts everywhere on all particles instantly. Similarly, in the many-universe interpretation (MWI) of QM, one seems to hang on to some vestige of *objective reality*, and moreover, the locality requirement is not really given up in MWI either.[13]

The question of (ζ) is a tricky conceptual one. If followed through, it may lead to a paradoxical situation. QM without measurement is meaningless. Measurement requires a measuring instrument. A measuring instrument has to be a *macro-level* thing that operates according to the laws of classical physics in order that it can be read reliably. However, according to QM, a measuring instrument *unavoidably* disturbs the object measured. Now, how reliable can a scale, that *itself* expands under heat, be for measuring the *rate of expansion* due to heat? It is like the problem of self-reference: how can I consistently say that 'I cannot speak at all'? One way out of it is to claim that 'it is illegitimate to translate *microscopic* uncertainty pertaining to the state of an atom to a *macroscopic* uncertainty pertaining to the measuring instrument'. Well, in that case, we need to be able to indicate where to draw a clear *line of demarcation* between macro- and micro-states so that we can tell exactly when the border has been crossed. Unfortunately, QM is silent here. Until we know exactly when and how the wave-function reduction takes over from the Schrödinger evolution, we do not have an unambiguous formulation of QM. Bell had strong reservations about QM on the grounds that it has no precise and deterministic prescription for predicting the outcome of a measurement.[14] Moreover, *if* it is really illegitimate to translate microscopic uncertainty to a macroscopic uncertainty, we naturally tend to ask whether the 'classical level' could be understood quantum mechanically. An answer in the negative entails two things: (i) QM is *incomplete*, at least in one sense (let us call it '*segmental* incompleteness') and (ii) if QM is correct only in its own domain, then we need two disparate sets of laws, one for quantum physics, the other for classical physics. Given our incomplete state of knowledge about physical nature, (ii) above is not a serious objection, but (i) is. So, Bohr rejects (i) and maintains that 'QM is as complete as possible'. John von Neumann even tried to give a formal proof of the 'completeness' of QM. In his book,[15] von Neumann had given a proof that no hidden parameters can exist. But in his proof he assumed that for all kinds of statistical assemblages, the laws of QM, expressed in terms of Ψ functions, are valid. If the indeterminism of QM is questioned (as HVTs do), von Neumann's assumption becomes suspect too. Thus, his 'proof' only shows that 'the assumption of hidden parameters is not compatible with universal validity of QM'.[16]. Von Neumann's proof, like some alleged proofs of Euclid's parallel postulate, involves *petitio*.[17] It needs to be pointed out that logically speaking, Bohr's claim that 'QM is *as complete as possible*' does *not* clash with Einstein's claim that QM ***is*** incomplete because it is quite plausible to surmise that no single common framework exists of which both Classical Physics and QM are components, each being valid in its own domain and further, that standard QM cannot undergo any piecemeal improvement without radically changing itself—lock, stock and barrel. Under such a semantic model, both Einstein's and QM's claims regarding the incompleteness/completeness of QM can be assigned the value 'True'/'False'. John von Neumann first proposed comparing projection operators in Hilbert space with propositions and their eigenvalues 1 and 0 with the truth values 'true' and 'false'.[18] This is comparable to Luckasiewiz's approach (in his book *Aristotle's Syllogistic*)[19] to Aristotle's problem about the '*Sea-battle tomorrow*'. As already indicated, I believe that the CP (classical physics)-QM divide is due to the problem of self-reference involved in linguistically formulating the theory of QM measurement in a consistent way *within* the linguistic framework $\mathbf{L}_Q$ of QM. It might be possible to do so only when we have a corresponding appropriate *meta-language* $\mathbf{L}_{MQ}$ (maybe somehow in a way analogous to assigning a unique Gödel number to a statement), so that it is possible to talk *about*, not just to talk *within*, the language $\mathbf{L}_Q$, in which QM theory itself is formulated. This, I would presume, is the real point that von Weizsäcker keeps insisting on, when he says that 'true logic is meta-logic'[20] or that 'true mathematics is meta-mathematics'.[21] He clearly suggests that true physics also needs to become *meta*-physics. 'It is tempting to say that "true mathematics is meta-mathematics" . . . mathematics seems to face problems that are . . . basically identical to those I am discussing here for physics'.[22] Therefore, to me, it looks reasonable to say that with the availability of such an L_{MQ}, QM can be expected to describe the *collapse of the wave-packet* without any problem. Until then, the objections stand.

As regards (ζ), it needs to be pointed out that in case of the measuring instrument (MI), along with the micro-level phenomena it measures, when both are considered part of one comprehensive QM framework, it leads to the vexed problem of self-reference. How can one consistently express one's *total inability* to speak English by using English itself to say so? No language L can be used without inconsistency as its own meta-language (i.e., an L_M), unless some clever way out of this difficulty can be worked out. This point was poignantly expressed by Wittgenstein.[23] In short, $L_Q \neq L_{MQ}$ in general, and we are yet to have a powerful meta-language L_{MQ} of the required type for QM. This very problem was dubbed by von Neumann 'the problem of infinite regress'. What it says is this: suppose that what QM says is literally true i.e., (i) there is no quantum world, no quantum Reality—only quantum potentialities, which become actualised only when observed, and (ii) by our first assumption, the MI is a part of the QM framework itself. If so, the MI has to be somehow observed *before* it can become actualised and do its job of QM measurement. So, when does the MI become a Reality? Only when observed by another instrument, and so on and on. It is not difficult to see that these are two different ways of expressing the problem of self-reference that crops up when the first option is taken. If on the other hand, the MI and the micro-level phenomena it measures are taken to belong to *two disparate* conceptual frameworks, then QM should better come up with satisfactory answers to the following two questions: (i) Where exactly lies the boundary line between the observed quantum system and the classical MI? (ii) Which equation of QM describes the collapse of the wave function? Conceptually linked with these is another question, viz., how far, and under what conditions (if any), can the framework of QM be extended to include macro-level phenomena? Penrose's complaint is that QM fails to give a proper recipe of how the quantum world merges with the classical.[24]

Let us consider (η) now. As I pointed out earlier, the metaphysical underpinning of the Copenhagen school of QM is positivist-operationalist in nature. It is also called 'black boxism' in philosophical parlance, because, according to its advocates, successful prediction is the *sole requirement* of a good scientific theory. The ability to predict is what a scientific explanation is all about. An otherwise elegant theory, if it fails to meet the predictability requirement, is not worth anything at all. Conceptual elegance may be kept waiting, but the requirement of the ability to predict cannot. A whole bunch of consequences follows from it:

(i) Any degree of predictability would be better than no predictability at all. So, if a probabilistic model of predictability works better than the causal model of prediction, drop the causal model. In short, for the Copenhagen school, only that which works is worth anything at all in science. Pragmatism reigns supreme. QM is still the best *prediction machine* available to us.

(ii) Accepting such a probabilistic model of prediction, in its turn, would require commitment to an *ensemble interpretation* of quantum phenomena.

(iii) Now, if one further assumes (as the advocates of QM do) that the QM framework is universally applicable, then a strictly atomistic worldview needs to be given up and needs to be replaced by some kind of holism.[25]

(iv) *If* we add to this Bohr's *prophetic* claim that QM is *as complete as possible* and no change is required in the existing structure of the theory, *then* it gives a stamp of *finality* to QM which smacks of a sort of intellectual arrogance and is counter to the more reasonable view that the growth of science is an open-ended process forever.

(v) Moreover, once we grant (i)–(iv) above, the Copenhagen version of QM would entail as its corollaries that (a) there is no quantum world, there are only quantum potentials which actualise as we observe them, (b) scientific explanation neither depends on nor methodologically requires that we keep looking for *that* enigmatic Reality which supposedly lies hidden behind the observed phenomena, and (c) in quantum theory, we are dealing with entities that go beyond the reach of pictorial representation.

Finally, regarding (θ), it should be noted that all the questions of the above type being *meta*-scientific in nature, there is no and cannot be any conceivable decisive experiment which can decide (in one way or another) between Einstein's 'realism' and the 'anti-realism' of QM. Moreover, the growth of science being an ever open-ended pursuit, any one or more of the underlying assumptions on which the *de*-ontologised anti-realism of QM stands can be (and actually has been) challenged in the course of the last two decades of research in the relevant areas. The relevance of these findings in the context of the Tagore-Einstein conversation will be discussed next.

Understanding the proper background of the Tagore-Einstein discussion

For a proper assessment of the significance and implications of the Tagore-Einstein conversation, it is essential that we look at it from the proper perspectives, both (i) scientific and (ii) socio-cultural. In this section, I will discuss them in that order.

Scientific background

First, we need to understand why a physicist of Einstein's stature felt an impassioned urge to discuss the issue with Tagore, a poet, not a scientist by any criterion. My answer to this is that Einstein's cherished worldview seemed somehow threatened by the anti-realist orientation of QM and its astounding success. A list of the triggering factors that heightened Einstein's threat-perception consists of the following:

(i) Bohr claimed that 'there is no quantum world but a realm of quantum *potentials* that become actualized only when observed', and Heisenberg claimed that in the QM framework, 'the notion of objective reality evaporates'. Whence, it follows as a corollary that the nature of Reality, whatever it be, is not something fixed, nor can it be pictorially depicted. 'We are beyond the reach of pictorial representation', Bohr is reported (by Rosenfeld) to have said to Einstein during a debate with him. To a staunch realist like Einstein, this might make QM look more like a fairy tale than objective science. This was a kind of *ontological* shock for him.

(ii) The spectacular predictive success of QM, coupled with the claim (of the Copenhagen school) that pragmatic success is the sole criterion of 'scientific explanation' and that it is both futile and pointless to look for a deeper explanation other than predictive power, posed a '*methodological* challenge' for Einstein.

(iii) Again, Bohr's explicit claim that his principle of complementarity is an *objective* and *the only possible* description because, 'the necessity of taking into account the conditions under which the experience is obtained calls directly for the complementary mode of description'[26] tends to suggest, in a way contrary to common sense, that Reality *must* always have a face hidden from us.

(iv) Furthermore, quantum entanglement, the so-called 'spooky action-at-a-distance' ('spookenhaft Fernwirkung', as Einstein put it) was like the last straw for Einstein's worldview, as it would force him to give up the *requirement of locality*. However, the requirement of locality, so dear to Einstein's heart, was neither intuitively obvious nor inviolable in principle. Besides this, nonlocality would undercut the dominant atomistic methodology of science because, atoms are, by definition, supposed to occupy *discrete* locations. All these factors (and similar other reasons *not* listed here) 'bristled' Einstein's instinct for physics. He was convinced that QM was not the last word, and it should be possible to give a deeper level of description of physical Reality that would be *complete* in his required sense. But why did Einstein's instinct for physics get bristled by these? Because they shook the very foundation of science on both ontological and methodological grounds and as such, was unsettling for our common-sense picture of the world. But why would a scientist, *qûa* a scientist, bother at all about saving a coarse common-sense picture of the world? Because, like Einstein, Whitehead believed that the 'very foundation of science is commonsense . . . you may correct it . . . but ultimately you must satisfy

> commonsense'.[27] Even Bohr had to admit that 'all experiments are to be described in classical terms', although he also kept insisting on the point that 'atomic objects and measuring instruments form an integral part of quantum phenomena'.[28] The primary methodological considerations that prompted Einstein to stick to classicist local realism may be called the 'principle of minimum mutilation' of an existing and useful explanatory framework. Clearly, QM did violate the principle of minimum mutilation, but Einstein's classicist local realism did not.

Irrespective of how justified Einstein's apprehensions about the onslaught of QM on the classical worldview of physics was, it would be safe to surmise that he was convinced that his common sense accommodating 'classicist' instinct for physics was basically right, and as such, he expected that any ordinary clear-headed person would unequivocally vouch for it. This explains why he would so impassionedly ask all visitors whether whether they *really* believed that the moon exists only when someone looks at it (like the quantum objects that are said to be actualised only when they are observed). Suppose that the answer he got fulfilled his expectation. Would it not be nice, a shot in the arm for his realist worldview, if a clear-thinking intellectual of Tagore's stature unequivocally supported his kind of worldview? Sure it would be. That is why perhaps Einstein was so eager to discuss the nature of Reality with Rabindranath Tagore, although Tagore was neither a physicist nor a trained scientist.

My above-mentioned hunch as to what prompted Einstein to be so earnestly engaged in a discussion with Tagore starts to look more plausible if we keep in mind the time sequence between the advent of QM (à la the Copenhagen school) in the late 1920s and Einstein's conversation with Tagore (1930) and the publication of the EPR paper (1935). Obviously, the EPR paper could not have been an overnight product. It must have been preceded by a prolonged period of serious intellectual self-searching. No wonder that the year of the Tagore-Einstein conversation regarding the nature of Reality (1930) is somewhere near the middle of the period (late 1920s and 1935).

Socio-cultural background

After discussing what 'scientific' considerations prompted Einstein to be so engrossed with the question of the nature of Reality, we may now take a look at the respective socio-cultural backgrounds of the two participants in the Tagore-Einstein discussion. It is needless to say that their socio-cultural backgrounds and moorings were as disparate as it could be. Einstein's metaphysical commitments (as can be gleaned from his responses to QM and his formulation of the EPR paper) were shaped by way of reaction to the newly developed QM worldview. I will first list the main distinctive features of the two views: classical physics and QM.

First, QM is non-atomist holism, while Einstein's view is atomistic. QM is a kind of non-atomist holism in a unique sense in the which atomism of the Jainas or the logical atomism of Wittgenstein as presented in his *Tractatus* (TLP) are atomistic, yet a form of holism at the same time.[29]

Second, QM supports nonlocality, while Einstein vouches for locality.

Third, according to QM, the elements of Reality have no definite features until observed; Einstein says that they do and *must* have definite features independent of any measurement. In other words, QM denies *counterfactual definiteness* (CFD); Einstein's denials demand CFD.

Fourth, QM drops causal determinism in favour of probabilistic prediction; Einstein adheres to the notion of linearly deterministic causality.

Fifth, according to QM, the nature of Reality is beyond pictorial representation; for Einstein, it can be pictorially represented.

Sixth, according to QM, only what can be observed (the phenomena alone) are real—there is no need to look for the 'real' beyond them; according to Einstein, the *true* nature of Reality lies hidden behind phenomena. So, we need to look deeper beyond and behind what appears to us.

Last, QM is 'integrative' in the sense that it refuses to separate the results of observation/what is observed from the process of observation. It proposes to include the measuring instrument and what it measures as integral parts of a single framework. An object is an object only for knowing subjects in the world. Man, the subject for whom the objects are objects, is part of the totality of objects. In short, QM rejects the notion of *pure objectivity* as a myth, whereas for Einstein, the notion of an objective reality, free of all traces of subjectivism in it, is the prime requirement of a sound scientific methodology. In the words of Einstein (to Tagore), 'I cannot prove that scientific truth must be conceived as a truth that is valid independent of humanity; but I believe it firmly'. Tagore responded, 'It is not difficult to imagine a mind to which the sequence of things happens not in space, but only in time like the sequence of notes in music'.

What are Tagore's metaphysical presuppositions vis-à-vis the above, and where do they lead him to as far as the question regarding the nature of Reality is concerned? Let us take up Tagore's metaphysical presuppositions first.

Tagore, steeped as he was in the monistic *Upanisadic* tradition by education and by family tradition, was a staunch supporter of non-atomic holism. Accordingly, he would view every discrete individual as an integral node of a cosmic network. For him, strictly speaking, the so-called discreteness of objects was unreal, a deceptive appearance. Again, the ultimate, all-comprehensive infinite Reality is 'One' and is technically called '*Brahman*' or the Cosmic consciousness. It is characterised as *bibhu* (ubiquitous). It is present always and everywhere—free of all space-time constraints. A view like this one may be construed as having a sort of *thematic affinity* to the notion of nonlocality. Tagore could not agree more with the view that the idea of an isolated object is an approximation and, according to quantum theory, a very bad one. There is no absolutely isolated object for us to know, and knowing that they do not exist in a strict sense, we correct this mistake describing them as interacting objects.[30] A.N. Whitehead put it thus:

'There is no society in isolation; electrons and atoms are societies. A society is an internally structured harmoniously interacting whole'.[31] Similar is Tagore's idea that every individual thing functions as an integral node of a cosmic network, so that '[w]hen one tugs at a single thing in Nature, He finds hitched to the rest of the World'. It expresses the same belief in *cosmic interrelatedness* as does the notion of quantum entanglement. Besides the monistic *Upanishadic* schools, similar ideas are also found in the pluralist-atomist school of Jainism. There it is dubbed 'pan-relational realism'. One sacred Jaina text, *Acārānga Sūtra*, expresses it thus: 'One who *really knows* the *true nature* of one single thing, knows the true nature of everything'.[32] Schrödinger once remarked that entanglement is not just one of the characteristic features of QM, it is *the* only characteristic feature of it.[33] As Kalidas Bhattacharya put it, 'in a sense we can say that belief in such cosmic interconnectedness is a pervasive feature of Indian philosophical tradition in general'.[34] Rabindranath grew up in this tradition. Moreover, true to the spirit of Indian tradition, he also believed that in the scheme of cosmic unity, the parts are as much dependent on the whole as the whole is on its parts. The two are so integral to each other that each yearns for the other in search of self-fulfilment. An individual consciousness and the cosmic consciousness blend in a way in which individual musical notes blend in a musical harmony. As Tagore put it in one of his poems: 'Infinity longs for a warm embrace from the finite, and finitude can't wait to lose itself in the bosom of the infinite'.

Now, in the light of the seven characteristic features of QM vis-à-vis Einstein's classical local realism, it is easy to see that Rabindranath's (and broadly speaking, of Indian philosophical tradition) metaphysical leaning about the nature of Reality has a deeper affinity to the QM position on most points, except the fourth and the sixth ones. Regarding causal determinism, the Indian position should be characterised as a form of inflexible and rigid consequentialism *without* any deterministic predictability pattern. This is so because the overarching pattern of cosmic regularity (called *Ṛta*) is a complex network in which an *indefinite number* of interwoven factors interacts to produce a resultant consequence which, though inescapable, can only admit of a *probabilistic* projection. However, if we could untangle each strand of the network that constitutes *Ṛta*, each of them would have ensured a strictly causal-deterministic, predictable consequence. Such an interpretation of cosmic regularity alone, I believe, can make normal sense of *karmaphalavāda* (every action has a certain consequence). Anyway, in keeping with the Indian tradition (including the Jainas and the Buddhists), Tagore's position in respect of the causality vs. probability view (i.e., 'classicist local realism' vs. QM, in short) of causal predictability would be a blend of the two (viz., causality and probability) and stay in the middle.

As regards the non-pictorial nature of Reality and its total lack of any *intrinsic* characterising feature, the nature of Reality is often indicated in Indian philosophy by words with *negative* connotations. For example, the

Buddhists call it *śūṇya* (pure emptiness); for the Jainas, it is *kevala* (pure *thisness*); for the non-dualist *Vedantins*, it is *nirguṇa* (uncharacterisable); and so on. Naturally, in this respect too, Tagore's position would resonate better with the QM view. One point on which Tagore's position would strongly differ from that of QM is regarding the need for looking beyond the phenomena. For Tagore, it is not only non-futile, contrary to Bohr's claim that it ***is*** so, it is also essential for grasping the true nature of Reality. Einstein would be happy to see that Tagore's position comes so close to his own. There is a school of philosophers called 'phenomenologists' who would put the question of 'what lies behind phenomena' on hold, advising us to *bracket* it out. Tagore's position in this respect was different from both phenomenology and QM. An urge to look beyond and behind the phenomena is in-built in us. It is an urge to *transcend* our finitude with a view to realise our spiritual *identity* with the cosmic consciousness through a sort of being 'attuned' with the infinite. This musical analogy colours Tagore's entire worldview.[35]

Naturally, the question of the man-universe relationship cropped up during his conversation with Einstein, too, especially when he was at pains to wean Einstein away from his crash objectivism (viz., mind-independent nature of Reality and of Truth) regarding the methodology of science. I *really* wonder whether Tagore succeeded in getting this idea (of 'a finite self merging with the Infinite into an organic one-ness') across to Einstein. A genuine, personal *realisation* of such one-ness (as distinct from an impersonal *theorisation* about the relation of identity), of what follows logically from a metaphysical notion of ultimate *non-dualism* of a finite locus of consciousness (an individual self) and the cosmic consciousness (= *Brahman*), is claimed to be *the* deepest *Upanishadic* truth viz., '*You* (= a finite locus of consciousness) are identical with *that* (the cosmic consciousness/Brahman) (*tat tvam asi*).

The following quotation from Arthur Eddington looks eerily similar to the *Upanishadic* view expounded above. To quote Eddington: 'Not once in the dim past, but continuously by conscious mind is the miracle of the Creation wrought'.[36] 'Where science has progressed the farthest, the mind has but regained from nature that which the mind has put into nature. . . . We have found a strange foot-print on the shores of the unknown. We have devised profound theories, one after another . . . And at last, we have succeeded in reconstructing the creature that made the foot-print. And Lo! It is our own'.[37]

As for me, I do not consider it a case of *accidental* chance coincidence or of mere lip service when an ardent supporter of QM and a direct student of Niels Bohr mentions *Brahman* and the corresponding doctrine of the non-dualist/monistic school of *Vedanta* in the context of QM. According to von Weizsäcker, 'mystical experience of unity is not an escape from rationality . . . but a consequence of understanding the nature of rationality'.[38]

However, for someone with a Judeo-Christian background like Einstein's (whose ideas about 'objective reality' and 'impersonal objectivity of truth' were no doubt coloured by the Judeo-Christian tradition via the staunch Cartesian view of consciousness-matter dualism), the very idea of *identity* of Man and God (= omniscient, omnipotent, infinite consciousness that oversees the cosmos) is an anathema. I wonder whether this might have resulted in what may be called, 'a conceptual incommensurability' between Tagore and Einstein. Anyway, for those who impute these qualities to God and consider Him to be a 'person', the very notion of God-Man identity is blasphemous—it would hardly be comprehensible to them. Needless to say, neither Einstein, Bohr nor Tagore had anything to do with the notion of such a personal god. For Einstein, 'The most beautiful and most profound emotion . . . is the sensation of the mystical. It is the source of all true science. To know what is impenetrable to us really exists . . . this feeling is at the centre of all true religiousness'.[39] The close affinity between science and mysticism was also emphasised by Bertrand Russell in his *Mysticism and Logic* .[40]

What would happen if . . .

I will conclude by raising some relevant and provocative methodological questions and by suggesting some highly speculative off-beat answers to them:

(i) I suggested earlier why Einstein was so very interested in carrying on the conversation with Tagore about 'objectivity' and 'truth'. Would the topic have been chosen if the threat from QM did not bristle his 'instinct' for physics? I doubt it. Had Einstein known of at least some of the later approaches to/interpretations of QM,[41] it may be, but *only* may be, that he would not have cared to put the question about the moon to visitors like Abraham Pais or about mind-independence of Truth and Reality to Tagore with the same kind of urgency, I would guess.

(ii) Had Tagore been successful in making Einstein drop his stark 'objectivism' regarding 'Truth' and Reality (in favour of allowing some observer-dependent factors in scientific theory construction), would that have compromised the methodological rigour of science? Is it possible at all to have a methodology of science that is, in principle at least, totally free of subjective (subjective ≠ whimsical/personal) elements? My answer to both of these questions is in the *negative*, as elaborated in (vi) below[42]. James Jeans is one of the prominent physicists of yesteryear who challenged the standard notion of scientific objectivity.[43]

(iii) Does any physicist of repute consider incorporating 'consciousness' as a plausible explanatory parameter? Yes, Wigner was one of those who did. He said, 'Through the creation of QM, the concept of consciousness came to the fore again: it was not possible to formulate the laws of QM in a fully consistent way without reference to the consciousness'. Wigner believed that the equations of QM would cease to be linear if consciousness was brought in.[44] It can be argued that Hardy's Paradox, viz. 'an act of observation affects the nature of the Reality that we seek to unearth', also supports Wigner's contention.[45]

(iv) I have already mentioned Tagore's 'attuning' model of the merging of finite consciousness with the infinite cosmic one like notes in a musical piece. This model can, with a slight twist, be used as a possible model for blending subjectivity of free choice with objectivity of its outcome. An analogy will make my point clear. Suppose that several radio stations A, B, C, D and so on are simultaneously broadcasting programmes at different frequencies and you have a tuner. You can tune into any one station *you* choose, but after you have tuned in, what you hear, namely the outcome (i.e., the particular piece of music) is not haphazardly picked up: It is objectively fixed. What Niels Bohr said in elaborating his notion of complementarity (viz., 'What you see depends on the experimental arrangement used') fits in naturally with our 'attuning' model *without implying* that the outcome (i.e., 'what you see') is random, has no objectivity and depends on the experimenter's whims.

(v) If Tagore, with his kind of metaphysical background, were a physicist, what kind of model/interpretation of QM would he prefer? With his strong leaning for 'cosmic connectedness' and his view that 'finite consciousness' and 'the cosmic consciousness' are complementary to each other, it would be reasonable to surmise that Tagore would prefer a *participatory* model of QM which would theoretically entail quantum entanglement without the baggage of 'branching universes' coming in toe.

(vi) Are all these speculations nothing but idle 'philosophising', venturing too far out into metaphysics, transgressing the legitimate boundary of physics? Not really. First, the line between physics and metaphysics is not fixed, and it is also too difficult for demarcation. Whether we want it or not, physics imperceptibly flows into metaphysics. David Bohm once lamented that 'philosophy', which means 'love of wisdom', does not find a serious place in physics. It cannot be overstated that modern science cannot maintain total immunity to metaphysics. No adequate *Weltanschauung* can be built on natural science alone.[46] Second, so far as bringing in 'consciousness' (as Wigner proposed) as a legitimate explanatory parameter in physics is concerned, we need to share the liberal attitude of Bell and maintain that it remains a *logical* possibility that it is the act of consciousness which is ultimately responsible for reduction of the wave packet. Our general reluctance to let consciousness play any active role in physical science is a crippling hangover from Cartesian dualism that proposed to keep consciousness and matter separated in two hermetically sealed compartments. I suspect that Einstein's penchant for 'objective realism' had its Cartesian roots. Once Cartesian dualism is rejected, it is quite plausible to regard consciousness and matter as the two sides of a single neutral stuff, not

> two utterly disparate substances. A complete picture of Reality must include consciousness. It would be unwise to deny to consciousness an active role in physics. 'Can we set up against our self-consciousness a "Nature" independent of it?' asked Max Planck.[47] In a similar vein, von Weizsäcker hypothesised, in a stark denial of Cartesian dualism, that '. . . my consciousness and my body are aspects of the *same* Reality',[48] as if it is consciousness acting in tandem with matter that creates the symphony called 'scientific objectivity'. Among others, Jeans, Planck, Schrödinger and de Broglie strongly emphasised such a point of view. It would also make the real point of Tagore's use of musical analogy ('music just can't be a *monophony* of a singer') more perspicuous.

Well, maybe such a view of letting consciousness in will win by simple headcount because it has an appealing romantic tenor. But actually, how reasonable is it, after all? Let us see. If we accept the Big Bang theory and take quantum entanglement seriously, *then* it follows that it is more reasonable to assume that consciousness and matter are like twins who share a common 'genetic code'. They are not like two unrelated antagonists. As John Gribbin put it, 'particles that were once together in an interaction, remain, in some sense, *part of a single system* which *responds together* to further interactions. Virtually everything we see and touch and feel is made up of collections of particles that have been involved in interaction with other particles right back through time to the Big Bang'. According to D' Espagnat, 'All those objects constitute an indivisible whole'.[49]

Moreover, research in the areas of cognitive science, our knowledge representation mechanism, the study of the patterns of brain rhythms (associated with different states of consciousness) etc. have given enough evidence to blast away what I called the *myth* of *pure objectivity* to which Einstein's kind of realists subscribe. We may collectively call these interrelated areas of study 'neo-epistemology'. My argument for what I call 'debunking the myth of pure objectivity', stripped to its barest outline, runs thus:

(i) Neuroscientists still do not fully understand how the brain manages to extract meaningful information: how the brain can efficiently synchronise the timing of spikes to encode information and rapidly solve difficult computational problems. This is because a group of spikes that fire almost at the same moment can carry much more information than can a group of comparable size that acts in an un-synchronised fashion. Researchers have developed 'neuro-morphic' electronic circuits that mimic this aspect of the brain's signalling network.[50]

(ii) Our in-built knowledge representation mechanism may process, interpret and also re-construct the sensory inputs to *create* the so-called 'objective world' as we know it. Such an 'objective' world need *not* bear any resemblance to sensory data/input (Cp. seeing a 3D world based on 2D retinal images).

(iii) This clearly shows the *inalienable* role of alternative *interpretative* ways of neural data processing that result in knowledge representation.

As a matter of fact, any conscious being can become a *knower* only when it comes saddled with its genetically built-in baggage of knowledge representation mechanism and is thus capable of playing an active role in a highly complex cognitive network which, so to say, forms a knowledge *Gestalt*.

(iv) Elmer Green of the Menninger Clinic in Kansas showed that certain *yogis* could control their brain processes and internal states (including working of the knowledge representation mechanism) merely through meditation and thought.

(v) Hence, by suitable extrapolations, neo-epistemology may also suggest the possibility of taking a scientifically *plausible theorisation* of the notion (contrary to Husserl's view) of a *pure object-free consciousness*, e.g., of *mystical* experience of the non-dualists, or, of a very similar cognitive state called *kalpanāpoḍham jñānam* by the Buddhists.

(vi) Scientists have also found that *when* meditators reach a state of deep awareness and internal mental serenity by their ability to control the pattern of δ rhythms of the brain, the *two hemispheres* of their brains—which ordinarily generate brain waves of *different* frequencies and amplitudes—*become synchronised*, and both hemispheres generate brain waves of the same frequency.

(vii) Let us now consider an interesting result in psychology: It has been experimentally shown that schizophrenia patients do *not* perceive some visual illusions, such as the tilt illusion, whereas under similar conditions, a normal person *does* and typically misjudges the tilt of a line.[51]

If so, who should we consider *normal*? Us who fall victims to the 'tilt illusion' or the schizophrenics who do not succumb to the 'tilt illusion' at all? Either way, it shows that such notions as of 'normalcy', 'objectivity' etc. are neither absolutely fixed nor properly standardised. Einstein's commonsense notion of *pure* objectivity cannot be an exception to it.

Since the Bohr-Einstein controversy ultimately boils down to the question, 'What is objective reality?', it follows that if my reconstruction of epistemology, (i.e. neo-epistemology) is sufficiently reasonable, then so-called 'objective reality' becomes primarily a 'subjective re-creation' out of whatever we experientially encounter in the world. Any knowledge of an external object has to be a queer *hybridisation* of 'objective' and 'subjective' elements. Such a creative blending of the 'given' (objective) with the 'imaginative' (subjective) is the life blood of poetic creation, the essence of which is suggested by the word 'poiesis' (the Greek 'ποίηος'). This should be a reason good enough to totally debunk the notion of 'pure objectivity'—a pristine kind of objectivity supposedly unsullied by any element of subjectivity at all.

These are just some of the relevant results (without the details) of studies in this area that are mentioned here in order to show that the notion of 'pure objectivity' is an untenable myth.[52]

All these (Big Bang, quantum entanglement and neo-epistemology) taken together may be said to have cleared the way for a legitimate entry of subjective factors in general and consciousness in particular in the scheme of scientific explanation.

Where do we stand, then, at the end?

Let us start by asking the following question: What is the best way of conceptualising the nature of Reality, assuming that what we have said so far is reasonable enough? Before we can give an answer, we need to offer some clarifications. A photographic snapshot and an oil painting portraiture may both properly portray a person's appearance. However, an oil painting leaves more room for artistic variations without necessarily failing to be *faithful* to the original.

A scientist, and even an ordinary knower, is no less of an artist than the painter. Similar is the case when it comes to a portrayal of what is called 'objective reality' by the scientists. Subjective-creative elements are always intermingled with the *given*, the data, so much so that it is not possible to tell them apart, even in principle. Pure objectivity is a fiction invented for methodological convenience and for bridging the gap between the common sense and scientific worldviews. What is considered 'objectively real' may change direction, like a weather vane, following the direction of a ruling paradigm in science. In scientific theorisation, objective reality functions like a *variable* in a formula, without having to have any fixed value in a context-invariant way. The notion of an objective reality is so abstract and intangible that it is difficult to define it precisely. Still, I venture to propose the following (purely provisional) definition of it: An experientially encountered element **X** is considered 'objectively real' in the context of a given theory T, provided that (i) **X** has some traceable link to the world of experience, (ii) **X** plays a non-ad hoc explanatory role in **T** and (iii) the shelf life of **X** is yet to expire. This expiry of shelf life may happen in many different ways; for example, due to lack of *simplicity*, more limited power of prediction-cum-unification etc. (Cp. Aristotle's *vis-à-vis* Newton's theory of gravitation, the Fitzerald contraction *vis-à-vis* special relativity in explaining the null result of the Michelson-Morley experiment). After the advent of Copernicus's theory, the shelf lives of 'epicycles' and the 'geocentric universe' expired. They are no longer considered 'real'; similarly, after the Special Theory of Relativity, the shelf life of the ether 'expired', and so on. The history of science is replete with examples of 'once-expired stocks' that were re-stocked again with a vengeance. None of these cases, however, will be discussed here.

So, between Einstein and Bohr, who wins? As I see it, it is not a question of winning or losing at all. Rather, we should keep in mind that '[a] clash of doctrines is not a disaster, but an opportunity'.[53] All of us should try to learn something from it. We have learnt how complex the deceptively simple question about the nature of Reality is. The methodological speculations I have suggested may look crazy, but I hope they would not look either indifferent or insensitive to the philosophical issues that underlie the debate concerning the nature of Reality.

Notes

1 Bohr, N., *Atomic Physics & Human Knowledge* (New York: Science Editions Inc.1961), pp. 1–3.
2 Jammer, M., *Concepts of Force* (Massachusetts: Harvard University Press, 1956).
3 Jeans, James, *Physics and Philosophy* (New York: Bobbs-Merrill Co., 1943), p. 200.
4 von Weizsäcker, C. F., *The Unity of Science* (Canada: Mc Grawhill-Ryerson, 1980), p. 290.
5 Sarkar, T. K., 'From logos to rationality in science', in *Facets of Rationality*, D. Andler, P. Banerjee, M. Chaudhury and O. Guillaume (eds.) (Delhi: Sage Publications, 1995).
6 Wittgenstein, L., *Tractatus Logico-Philosophicus*-5.6 (London: Routledge & Kegan Paul, 1969).
7 Einstein, A., Podolsky, B. and Rosen, N., 'Can quantum-mechanical description of physical reality be considered complete?', *Physical Review* 47 (1935), pp. 777–80.
8 Bohr, N., 'Can quantum-mechanical description of physical reality be considered complete?', *Physical Review* 48 (1935), pp. 696–702.
9 Fine, A., 'Einstein's critique of quantum theory' in *The Shaky Game: Einstein, Realism, and the Quantum Theory,* 2nd edn. (Chicago: University of Chigaco Press, 1996).
10 Einstein, A., 'Autobiographical notes', in *Albert Einstein: Philosopher-Scientist*, A. P. Schilpp (ed.) (New York: Biography of Living Philosophers, Harper Torchbooks, 1949), pp. 83–7.
11 Bell, J. S., 'On the Einstein Podolsky Rosen Paradox', Physics 1(3) (1964), pp. 195–200.
12 See Bohm, D. and Hiley, B. J., *The Undivided Universe: An Ontological Interpretation of Quantum Theory* (London: Routledge, 1993).
13 Everett, H., ' "Relative state" formulation of quantum mechanics', *Reviews of Modern Physics* 29 (1957), pp. 141–53; Wheeler, J. A., 'Assessment of Everett's "relative state" formulation of quantum theory', *Reviews of Modern Physics* 29 (1957), pp. 463–65; see also Sudbery, A. in this volume.
14 Bell, J. S., 'Against measurement', in *Speakable and Unspeakable in Quantum Mechanics* (Cambridge: Cambridge University Press, 1987), pp. 213–31.
15 von Neumann, J., *Mathematical Foundations of Quantum Mechanics* (Princeton, NJ: Princeton University Press, 1955), from the German edition (Grundlagen . . ., 1932), pp. 160–73.
16 Reichenbach, *Philosophical Foundations of Quantum Mechanics*, p. 14.
17 See Sarkar, T. K., *Emergence, Complexity and Reductionism*, Unpublished Ph.D. Dissertation, University of Waterloo, Canada, 1978.
18 von Weizsäcker, *The Unity of Science*, p. 196.
19 Cp. Aristotle's problem about the 'Sea-battle tomorrow' in Lukasiewicz, J., *Aristotle's Syllogistic from the Standpoint of Modern Formal Logic,* (Oxford: Oxford University Press, 1957), pp. 155–56.

20 von Weizsäcker, *The Unity of Science*, p. 197.
21 Ibid., p. 203.
22 Ibid.
23 Wittgenstein, *Tractatus Logico-Philosophicus*, 2.172, 2.173, 2.174.
24 Penrose, R., *The Emperor's New Mind* (Oxford: Oxford University Press, 1999).
25 Bohr, *Atomic Physics & Human Knowledge*, Lectures of 1937 & 1939.
26 Ibid.
27 Whitehead, A. N., 'Science is no fairy tale', *The Interpretations of Science*, A. H. Johnson (ed.) (New York: Bobbs-Merrill Co., 1961), pp. xii–xiii.
28 Bohr, *Atomic Physics & Human Knowledge*, p. 2. It should be noted that Bohr's non-atomist holism is unique in the sense in which atomism of the Jainas or the logical atomism of Wittgenstein (as presented in TLP) are atomistic, yet a form of holism at the same time.
29 In Sarkar, T. K., Sanyal, I. and Moitra, S. (eds.), *Wittgenstein* (in Bengali) (Kolkata: Allied Publishers, 1998), pp. 81–94; also Sarkar, P. (ed.), *Studies in Epistemology: Indian Perspectives,* (Calcutta: University of Calcutta in association with RADIANCE, 2009), Chapter 3.
30 von Weizsäcker, *The Unity of Science*, pp. 190–215.
31 Palter, R. M., *Whitehead's Philosophy of Science* (Chicago: University of Chicago Press, 1960), pp. 216–17.
32 Suri, Haribhadra, *Sad-darsana samuccyaya*, L. Suali (ed.) (Calcutta: The Asiatic Society, 1986), p. 222. Translation mine.
33 Schrödinger, E., 'Discussion of Probability Relations between Separated Systems' in *Mathematical Proceedings of the Cambridge Philosophical Society* 31(4), pp. 555–63.
34 Bhattacharya, Kalidas, *Bharatiya samskiriti o anekanta vedanta* (in Bengali) (Burdwan: University of Burdwan, 1982).
35 Tagore's Book of Songs *Geetavitaan* (in Bengali), especially songs numbered 335, 347, 356, 363. Two relevant books on the Tagore-Einstein conversation are Goslin, D., *Einstein Met Tagore* (New York: Routledge, 2007). Topics in thus book include: pp. 130–32 on Tagore's major themes; pp. 161–65 on Nature of Reality. Goslin's book also has a chapter on Sri Aurobindo. Another earlier but helpful book on the topic is by Frank, Philip, *Einstein: His Life and Time* (Camridge, MA: Da Capo Press, 2002, reprint of the 1947 edition), with pp. 113–43, on Einstein and Rabindranath.
36 Eddington, A. S., *The Nature of the Physical World* (Ann Arbor, MI: University of Michigan Press, 1978), p. 24.
37 Eddington, A. S., *Space, Time and Gravitation* (Cambridge: Cambridge University Press, 1978), pp. 200–20.
38 von Weizsäcker, C. F., *The Unity of Science*, p. 391.
39 In Barnett, L., *The Universe and Dr. Einstein* (New York: Bantam Books, 1979), p. 108.
40 Russell, B., *Mysticism and Logic* (London: Unwin Books, 1963), p. 9.
41 J. S. Bell proposed that the idea of super-determinism could explain the result of two particle experiments without postulating a signalling speed > c only if the two particles have been predetermined by causes reaching both from an earlier time. Bell's super-determinism denies freedom of choice to the experimenters (London: *BBC Talk* organised by P. Davies and J. R. Brown, 1980). Later, John Conway and S. Kochen claimed in their paper on 'Free Will Theorem' that 'free choice' of the experimenters would require that atoms have freedom.
42 Sarkar, T. K., 'Epistemology: Limits and transcendence', Lecture at North Bengal University, December 2014 (to be published), and 'Theory of knowledge: A neo-epistemic approach', Lecture at the University of Calcutta, April 2015.

43 Jeans, J., *Physics and Philosophy* (New York: Bobbs-Merrill Co., 1943), p. 200.
44 Quoted in Venkataraman, G., *Quantum Revolution III: What Is Reality?*, Vignettes in Physics (Hyderabad: University Press, 1994), p. 72.
45 Lucien Hardy, 'Quantum mechanics, local realistic theories and Lorentz invariant localistic theories', *Physical Review Letters* 68(20) (1992), pp. 2981–84, and 'Non-locality of two particles without inequalities for almost all entangled states', *Physical Review Letters* 71(11) (1993), pp. 1665–68. Also, Hardy's Paradox, *New Journal of Physics*, March 2014.
46 Sarkar, T. K., *Knowledge, Truth and Justification* (Kolkata: Allied Publishers, 1992), pp. 169–73. Also, Sarkar, T. K., 'Philosophy, science and philosophy of science' in *History of Science and Philosophy of Science*, P. K. Sengupta (ed.) Vol. XIII, *Part-7 of Philosophy and Culture in Indian Civilization, PHISPC* (Delhi: Pearson, 2012), pp. 3–29.
47 Planck, M., *A Survey of Physical Theory* (New York: Dover Publications, 1960), p. 53.
48 '. . . my consciousness and my body are aspects of the *same* reality', von Weizsäcker, C. F., *The Unity of Science*, p. 254. Also see Whitehead, A. N., 'Science is no fairy tale', p. xxxix.
49 D'Espagnat, B., quotation taken from Venkataraman, G., *What Is Reality?, Quantum Revolution III, Vignettes in Physics,* (Hyderabad: Universities Press (India), Ltd., 1994), p. 67.
50 Boahen, Kwabena, 'Neuromorphic microchips', *Scientific American* 292(5) (May 2005), pp. 56–63.
51 Sejnowski, T. and Delbruck, Tobi, 'The language of the brain', *Scientific American* 307(4) (October, 2012), pp. 54–9. Also, see Dutton, Kevin, 'Wisdom of the psychopaths', *Reviewed in Scientific American* 307(4) (October 2012), pp. 76–9.
52 Sarkar, 'Epistemology', Note 42.
53 Whitehead, 'Science is no fairy tale', p. xxxi.

7 Physical reality and the unobservables of physical nature

C. S. Unnikrishnan

Introduction and scope

The fundamental physical theories that interpret and explain the behaviour of matter in nature are dependent on several unobservables and insensibles in their construction. Entities like fields, wave functions and even space and time are all unobservables, except as manifestations of material existence and behaviour. There is thus an obvious difference of degree and meaning between the reality associated with these unobservable theoretical entities and that of perceptible matter. The success of the physical theory is often taken as evidence for the physical reality of such unobservables. While a rigorous natural philosophy will not be able to support or approve their reality with the same vigour and conviction as it might defend the reality of matter, there does not seem to be a way of avoiding such unobservables if we have to construct theories. Though there is compatibility and consistency between observables and unobservables in most of classical physics, apparent conflicts and dissonance arise when microscopic physics is to be understood with a satisfactory theory. There are even observational consequences highlighting such conflicts when cosmology and the dynamics of the universe are included into the larger physical framework.

In this chapter, I examine the nature of physical reality in the context of unavoidable unobservables in physics[1,2] and discuss some examples. For the purpose of this discussion, I will work with a definition of an unobservable (in physics) as a quantitative entity, *created* and mathematically representable in relation to a theory of sensible matter and its behaviour, but whose ontological status cannot be established nor demonstrated directly or by deduction employing methods usually used for material entities. Thus, an unobservable cannot be proved to exist by tangible sensing with any conceivable observational device or even by a logical argument relying on empirical evidence, with a degree of conviction nearing that of the reality of material existence that we normally admit in the context of physical science. Another description, in line with the history of the use of such unobservables, would be as entities created or postulated, in the context of the physical theory, which we are willing to believe as real but cannot be proved to exist or otherwise by any empirical means known to science. In this sense, such unobservables belong to a projected reality involving the human mind. Clearly, we do not have a firm definition to go by, but I hope that examples can serve as clear indicators.

I do not plan to discuss the issue of realism in the sense it is usually debated on, where questions or doubts are raised whether what is perceived implies an underlying objective reality and existence. For the purpose of our discussion, such debate is a distraction. In fact, the view taken here is that the issue of real existence independent of perception of the human being is irrelevant for human endeavours like science. All that is required for science is the internal consistency and stability of perceived and observed patterns of nature. While it is reasonable to extrapolate that to an objective reality underlying perceived phenomena, there is no logical necessity that we establish that reality before building theories of phenomena. This is not a point to dwell on within the scope of this chapter.

One of the features of science, especially physical science, is a powerful urge to construct theories of phenomena, instead of just cataloging and classifying phenomena and their spatio-temporal patterns. Theories go beyond the mathematical modelling of patterns. For example, the mathematical and geometrical description of planetary motions, which of course contains some aspects of theoretical construction of a mathematical nature, attains the status of a physical theory only when completed with an explanatory description involving the massive sun, gravity, inverse square law etc. The ability of the theory to include a reasonable explanatory power, addressing the questions of 'why' and not just 'how' is important in natural philosophy and physics.

As soon as we embark on this programme of theoretical description of observed phenomena, we face the problem of having to conceive and include entities that are not directly sensed, but could reasonably be argued to 'exist' in nature. More often than not, such entities remain unobservable, with only plausibility and consistency arguments in support of their reality. However, they are, by design, fully representable mathematically and they bear definite relations to observable quantities and other elements in the theory. Often, the motivation for introducing an unobservable is a strong commitment to causal development of the physical world through a series of cause-effect relations. The justification is made stronger subsequently if there is relatively long-lasting success and consistency of demonstrating the cause-effect relations in phenomena relying on such an entity. A concrete example, the theory of planetary motions, can clarify the essence of such a structure of the physical theory.

Physical theory with an unobservable: an example

There were pre-Keplerian descriptions of planetary motions with supporting geometrical construction that did well in the construction and prediction of the ephemeris. Kepler's model of planetary motions based on a compact set of statements or 'laws' improved the precision of dealing with planetary trajectories, but still did not provide an answer to why planets move in elliptical orbits in precisely the way they do. That explanation had to wait for Newton's theory of gravitation and his identifying the long-range gravitational action as responsible for holding the planets in their orbits. All of Kepler's laws emerged as consequences of the underlying physical theory advanced by Newton, with the inverse-square law for gravity and the conservation of angular momentum. But the transition from mere laws to a physical theory also brought in the need to hypothesise the phenomenon of non-contact gravitational action that can 'hold' an object at a distance under a force without any contact. The disbelief about such a possibility is cured by postulating a field of gravity (which was done long after Newton's time), continuous and contiguous from one material body to another, acting like an invisible and insensible yet 'real' entity. We clearly see in this example the necessity and role of an unobservable in a fundamental physical theory. Faraday is rightly credited with firmly presenting the concept of the ethereal fields of electricity and magnetism and with asserting and defending their 'reality'. Phenomena based on electromagnetic radiation are now considered sufficient proof for their physical reality. Any attempt to avoid such pervading fields makes the theory very complicated and subject to criticisms of non-local action at a distance. In any case, historically, a 'field' has become an integral part of physical theories in spite of our inability to really prove its reality with the same degree of conviction as the reality of matter.

Fields, sources, space and time

The concept of 'fields' of various kinds dominates physical theory and modern physics, and fields are essential to both classical and quantum theories. However, the classical fields we are familiar with are somehow in good harmony with our notions of plausible reality, perhaps because such fields were postulated in an intuitive way in the context of electromagnetism and gravity. They are in fact natural extensions of sensible material fields, like the velocity or density at different points of a flowing liquid or the temperature at different locations in space. These material fields are related to some physical property of continuously distributed matter. Hence, once the reality of matter is accepted, the reality of such a field is not in serious doubt.

In contrast, a magnetic field or a gravitational field is characteristically different. They certainly need a material source, like a current or a massive body, but fields are posited to exist in regions where there is no source. The field is the physical device that connects one material object to another spatially separated one, and the field is the agent of the interaction and the evolution from cause to effect. Such fields are 'observed' only through the dynamical behaviour of a material body—a test particle, as it is often described—in the presence of some spatially separated source matter. Here, we see a secondary layer of reality that can be questioned into doubt. Fields like the electric and magnetic fields and the gravitational field are theoretical constructs, posited to exist to simplify both the mathematical structure of the theory and to preserve an intuitive understanding of physical phenomena as due to contiguous cause-effect relations in space and time. One of the implied tenets of physical sciences is the need to empirically prove the existence of an entity as 'physically' real. However, this cannot be insisted on in the case of such fields. They remain unobservable in the sense of 'observability' as applied to the reality of matter, while being consistent with the hypothesis of their reality in physical phenomena.

Einstein had discussed the relation between matter, fields and space in the context of the development of the theories of relativity.[3,4] The point of view expressed there is that fields are physical states of space, modifiable by the presence of material sources. (In fact, this was his motivation for constructing unified field theories where there is only one structure of space when both gravity and electromagnetism are considered together.) Though he gave importance to sensible experience as the true basis for concepts with ontological content, one sees in his writings the preference to give superior status relative to matter to the geometric empty space. (In his article 'Physics and Reality' (ref. 3), he compared his field equations to a building with one of its wings made of fine marble, standing for geometry of space-time, and the other made of low grade wood, meaning matter. It is iron that he has to refer to two forms of matter itself to make this distinction.)

There is a strong relation between the familiar fields and the state of motion of matter. While an electric field represents the action between static charges, the magnetic field represents the action of charges in motion, or 'currents'. When motion ceases, the magnetic fields disappear. Also, curiously, only moving charges can 'see' a magnetic field. There are similar features in the modern theory of gravitation. Hence, the reality of such fields is dependent on the state of the motion of the observer or that of the device used by the observer to 'sense' the field. This feature illustrates the subtle aspects of dealing with the issue of reality of even familiar entities in the physical world.

At least in one context, the hypothesis of a 'field' seems to be more than a convenience. That is the phenomenon of radiation. Light can in principle be thought of as the response of charged particles in the eye to another charged particle moving periodically somewhere else far away. In this view, there are only (charged) material particles. However, thinking of light as a propagating field in its own right and reality, separated from the source or a potential detector, is considered to be essential for a consistent description. Indeed, light of that kind is considered as particles in their own right, called photons. Therefore, understanding the reality of photons is, and will remain, crucial for understanding a whole lot of ontology in physical theory. One important aspect to note in the case of such 'radiation fields' is that they are retarded in time—there is a time delay that increases with distance between the cause (motion of the source) and the effect (response of a detector). So, perception itself is not perception of 'now' in an absolute sense. This is the basis of the denial of the absolute simultaneity of events in relativity theories. We will come back to a discussion of the field of electromagnetism when we discuss the 'physical properties' of the unobservable vacuum or emptiness.

There are two fundamental 'fields' used in physics, with progressively more physical roles ascribed to them as the theory progressed. These are space and time. While Newtonian physics needed space and time with some precision in their definition distilled from the common sense use of these concepts, modern relativistic physics treats these as dynamical fields in a sense similar to the way electromagnetic fields are treated. A fuller discussion of this is not intended here. However, we will occasionally comment on the status of space and time, or space-time as it is referred to in relativistic physics, as physical fields. A discussion of the reality of space and time is fraught with many dangers of logical and philosophical complexities. Simply put, a concept like time does not make sense without matter. All of our notions of time are changes in material configurations. While space without matter is 'imaginable' in a way, even those notions of spatial extent etc. are generalised from our familiarity with material reality. Quoting from Einstein's forward to Jammer's book[5] *Concepts of Space*, '. . . two concepts of space may be contrasted as follows: (a) space as positional quantity of the world of material objects; (b) space as container of all material objects. In case (a), space without a material object is inconceivable. In case (b), a material object can only be conceived as existing in space; space then appears as a reality which in a certain sense is superior to the material world. Both space concepts are free creations of the human imagination, means devised for easier comprehension of our sense experience'.

Space and time are supreme and primary unobservables in physics, without any reality independent of matter, notwithstanding their modern status as the dynamical arena of gravitational effects as described in Einstein's general theory of relativity. However, they are essential for even starting to imagine about a physical theory, let alone construct it. The situation is different from the notions of electromagnetic fields or the gravitational field because space and time are continua without material sources. Their existence and reality are the most primitively entrenched notions in our minds, and yet they are primary examples of unobservables in physics. The difficulty of incorporating space and time into the physical theory without referring to matter has been realised from the very early times of mathematically precise physical theory, as evident in Newton's *Principia*:[6]

> Absolute space, in its own nature, without relation to anything external remains always similar and immovable. Relative space is some movable dimension or measure of the absolute spaces, which our senses determine by its position to bodies, and which is commonly taken for immovable space.

> But because, the parts of space cannot be seen, or distinguished from one another by our senses, therefore in their stead we use sensible measures of them. For from the positions and distances of things from anybody considered as immovable, we define all places, and then with respect to such places we estimate all motions . . .

As it is well known, a well-reasoned critical thesis on space without material references as the arena of physical effects had to wait till Mach's critique on Newton.[7] Unfortunately, modern physics went retrograde on this aspect, denying Machian insights, and we will see later that we are destined to pay the heavy price for this.

Quantum physics and its unobservables

Twentieth-century physics brought in new notions of particles, trajectories and causal relations for dynamics (also called 'laws' of physics), which dissolved or even denied the earlier established notions of identity, individuality, distinguishability and localisability in space and time: hence, a new notion of reality was to be developed.

Most of modern discussions on reality in physics are linked to the microscopic physics described by the quantum theory or quantum mechanics (QM).[8]A major instigator for this situation was one of the prominent contributors to the theory, Einstein himself, who asked the most pertinent questions about the representation of physical reality in QM.[9] It is in the context of QM that one first encounters the necessity to represent the dynamics of a particle with an entity that has properties that are familiar from classical physics but has no tangible existence in real space—a wave to which properties like wavelength can be ascribed in direct relation to mechanical properties of the particle, but that has no real existence as a physical wave that propagates in real space. The 'wave function' in QM is an unobservable field without a source. It is inseparable from the existence of the material particle it represents; yet it is not the material particle. It holds in its description all the physical properties of the material particle, like charge, mass etc. and interacts with other material sources in a conventional way through their 'fields', but it has no tangible traces in real space. It can be split into multiple parts by simply providing possibilities of splitting, with each part holding physical properties of the material particle it represents *in entirety*, yet the different parts do not interact with each other in any way. In short, no consistent, real, physical existence can be ascribed to a wave function. However, the entire description of microscopic physics today depends on the wave function representing all possible physical behaviour of the particles as if it exists in some way in space, feeling the external world through interactions.

The only tangible relation between the wave function and observable entities is statistical—the absolute square of the wave function is related to the probabilities of relevant observations. Every observation in general 'resets' the wave function, so to speak, and a new evolution starts. Perhaps the QM wave function is the only unobservable in physical theory that has not been assimilated into the common sense of the physicist.

It is tempting to commit the mistake of identifying the wave function with the particle in many situations, loosely calling it 'matter wave', especially in those situations where one deals with just one particle. The underlying representational feature is the familiar wave-particle duality. Indeed, this was the kind of intuition that people tried to cultivate and discuss in the early days of QM. This programme fails in the details, and no tangible reality or even a describable nature of existence could be assigned to the wave function. However, even today, many discussions, especially those dealing with quantum interferometry involving particles (called matter-wave interferometry), try to describe the underlying physical phenomena as if the wave function has a real physical existence in space, which is generally an inaccurate and inconsistent notion.

Some problems arising from the unobservables of QM

Before we discuss some problems that arise from the use of the unobservable wave functions in QM, I want to stress that there is no known conflict or inconsistency between QM and the assertion that material objects have an objective reality of existence, in the sense of their possessing some physical properties as well as occupying some region of space even before an act of observation has been done. Indeed, writing a wave function in QM assumes the physical reality of the material system with some physical properties, like mass, spin etc. Therefore, commonly found statements like 'an atom or even a stone has no properties prior to observation', or 'the moon exists only when it is observed'[10] etc. are not implications of QM. However, what is true in the QM representation employing a wave function is that the representation cannot be directly interpreted as providing a correspondence to reality in any familiar terms. Thus, the wave function does not hold in itself information of exact position, exact velocity, exact components of angular momentum etc. The wave function represents the physical quantities without 'possessing' the physical properties of the particle, like energy or momentum. Yet, it holds the exact probabilities for these quantities taking specific values or ranges of values on observation. Also, the wave function is a divisible entity, much like a real wave, whereas the underlying material system is not. What is observed or sensed is only the material system, and then the wave function transforms suitably to represent the new information of observation and many of its multiple parts disappear from consideration. No one is competent to say whether it disappears as a physical entity because its ontology is not known yet. The correspondence between the matter and wave function in the theory is such that only the probabilities of realisation of particular values of physical quantities are transformed or renewed, and the physical system and the totality of its properties (like total energy) remains conserved, whereas the wave function itself is not. There are also situations where there is a QM representation of the physical properties of a material system consisting of many (possibly two) particles, but there is no QM description at all for any of the particles taken individually. Therefore, the ontological status of the wave function and its reality as a physical entity etc. are not properly understood.

Yet, there is something physically real about the wave function, as can be seen from a simple consideration. Figure 7.1 represents a situation of 'interference', familiar in the case of light or other waves.[11] Essentially, the amount of light that exits from port 1 or 2 depends on the physical properties, like the refractive index and path length, of both paths, A and B. Variation of the difference in path lengths changes periodically the amount of light from either port between a maximum and zero (or a minimum close to zero).

Therefore, one can have a situation where the entire light exits one of the ports with the other remaining totally dark. Small changes in the path length then can change the situation and reverse it. The bizarre possibility in QM is that similar interference happens for material particles sent in one by one into such a device, as if each particle has an associated wave. Irrespective of what the physical picture is, the fact remains that the port through which the particle exits can be controlled by small changes in either or both of the possible paths, and the total change from port 1 to port 2 can be done deterministically by changing one or both of the path lengths by a physical movement of the material elements like the mirrors. It is then very difficult to maintain the view that there was nothing physically real in those paths, even though moving some mirrors in those paths changed a physical outcome. If one accepts that physical changes could happen only because the

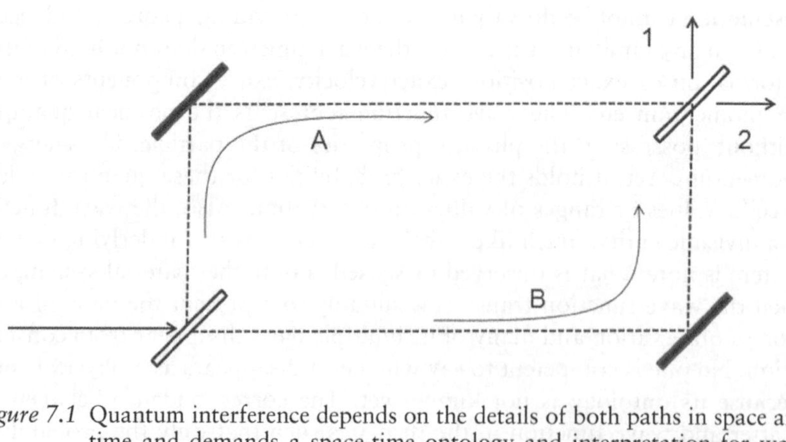

Figure 7.1 Quantum interference depends on the details of both paths in space and time and demands a space-time ontology and interpretation for wave functions as well.

mirror interacted with and affected something real in space, then one will also have to accept that the entity was present in both paths simultaneously, even though the material particle could not exist in both paths simultaneously. Further, one will have to face the consequence that if the particle was looked for and found in one of the paths before the exit ports, then the wave function corresponding to the path where the particle was not found has to disappear from the theory and by implication, from any possible physical existence—that step is required to preserve the relation of the wave function to probabilities of observations.

From classical reality to quantum unreality

A classical particle sourced with some random energy *E* will have some energy, momentum etc., the definite values of which can be known only after a measurement, but we do not contest that the particle possesses some specific energy even before measurement which shows up when an actual observation is made. So, the energy state of the particle might be specified as (E_a or E_b or E_c or . . . E_x) with corresponding probabilities (p_a, p_b . . ., p_x). In this case, we believe that a particle does possess one of these energies between the source and the detection, even though we specified the physical states with a possible set of energies and probabilities. In QM, the physical state that is not specified like this though this classical statistical feature can also be included in the specification. In the simplest cases of dealing with a particle, it may be that the state is specified through a wave function as $S = S(E_a) + S(E_b) + . . . + S(E_x)$, called a superposition of states with different energies. In this case, the square of each term encodes also the probability of observing the particle with that particular energy. However, it will be wrong to interpret that the particle has one of these energies before an observation. It is also wrong to say that it had all these energies at the same time. Similarly, if a particle is actually observed in a small region of space, QM does not allow any claim that it was somewhere nearby just before the observation. The '+' sign here does not correspond to the 'or' in the specification of the classical state, nor does it represent our lack of the precise knowledge of the physical state. Any translation of the mathematical state into any other language does not seem to capture its meaning. The unobservable in the theory is also untranslatable. This is perhaps the point in the quantum description where the familiar concepts of reality lose their anchor. The best one might do is to say that each term $S(E_x)$ represents some kind of wave with energy E_x, and we have a superposition of such waves as the representation. But this does not help much because the observations are of the material particle and will result in some specific value of energy, and then we are forced to say that all other waves in the superposition simply disappeared after the observation. In spite of this, the particle has a definite QM physical state, S. There is no confusion on this fact. So, within QM, a physical state need not be restricted to a state with specific values of each physical quantity. There is a well-defined QM state, even though we are not able to answer questions like what is the energy of the particle.

The main features of the QM representation can be discussed in a simple manner. The single characteristic feature is the superposition of representations of physical states and the resulting 'incompatibility' or mutual uncertainty relation between measurements of some pairs of quantities. There are many such 'incompatible' physical quantities in the quantum world, and examples are {position and momentum}, (spin in the x-direction and spin in the y-direction} etc. In fact, this incompatibility between certain pairs of observable properties, contained in the possibility of superposition, is what distinguishes QM from classical physics.

If a physical system has a measurable property A, perhaps with just two possibilities of outcomes labeled H and L, QM represents this as two possible states of being (H) and (L) with associated measurement values and asserts that any linear combination of the states (H) and (L) is also a valid and possible state. Obviously, such a linear combination, say $(H) + (L) = (U)$, cannot give a definite value of H on all measurements. (Here we have ignored writing the numerical factors because they are not relevant for the discussion. However, they are important because the square of the coefficient is the probability to find the system in that particular state, if observed.) If the superposition is in equal proportion, the theory associates equal probabilities for the observation of H and L and in general, the square of the coefficients of the superposition determines the associated probabilities.

Consider the example of an atom, which can either be in the state (High) or (Low). Hence, a possible QM state of the atom could be written as the superposition (High) + (Low). Though one might be perturbed about the meaning of the state of being like (High) + (Low), once it is understood that it is a representation with no implication that the atom is both High and Low at the same time, one can progress further. All we need to accept at this stage is that the state (High) + (Low) (or (H) + (L) for short) represents a valid physical state, even if we are not able to immediately 'understand' what it means. In some cases, there is a straightforward interpretation for such states, which is comforting. For example, if the state labeled (+Z) represents a state of the spin of an electron being along the Z direction (in an x, y, z coordinate system) and the state (–Z) represents it being in the negative Z direction, the state (+Z) + (–Z) is in fact the state in which the spin is along the X direction. In the hypothetical case we were discussing, the state (H) + (L) might be a definite state for some other property, possibly dichotic, with symbolic values 'Up' and 'Down'. The scope of QM is contained in the representation that allows superposition where combinations of two states of one property express also two states of some other property. For example, symbolically, the two states 'Up' and 'Down' can be expressed as superposition of (H) and (L): $(U) = (H) + (L)$ and $(D) = (H) – (L)$. Adding complex numbers to the scenario allows more measurable properties and states to be expressed.

This simple representational feature contains the essence of QM reality. We see the uncertainty principle in action right away. If the physical system is prepared in the physical state (U), for example, but we decide to observe with a suitable apparatus to see whether the atom is in state (H) or (L), we will get random results with equal probability because the state (U) is also the state $(H) + (L)$. Also, a state prepared as (H) is really the combination $(U) + (D)$ (with a multiplying numerical factor, which we ignored), and the uncertainty is reciprocal. Clearly, QM uncertainty is not because the source randomly sends out the states (U) and (D) or (H) and (L). The representation of physical states with superposition precludes the naïve reduction of QM to a classical statistical theory.

QM explicitly prohibits any interpretation that if 'High' is found on observation, then the state was necessarily 'High' just a moment before the observation. It could have been (U) or (L) or some more general superposition of the two. That is the crucial difference between classical statistical observations and the ones in QM. Therefore, in QM, there is a *collapse* of the state upon observation, a sudden change from a general state of superposition to a reduced state. This is not a sudden change in just our knowledge of the physical state of the system (as it would happen in classical physics as well), but it is a change of the physical state itself, within the QM formulation.

Unfortunately, no real progress has been made on clarifying the associated ontology, or the lack of it, in the past 80 years or so, throughout the entire history of QM. If at all, the puzzles and surprise have become stronger due to the many interesting experiments that have been performed (ref. 11).

The case of two particles and two properties

The representation of the physical state involving two particles, each possessing two physical properties that cannot be simultaneously specified to arbitrary accuracy, reveals some subtle aspects of reality in the context of quantum mechanical descriptions of the microscopic world. It is this kind of an example (with position and momentum as the physical variables) that Einstein and his collaborators discussed in the Einstein, Podolsky and Rosen (EPR) argument in 1935, that the QM representation is not a faithful representation of the real physical state.[12] Almost the entire modern discussion on reality in the quantum world is in the context of such examples and the EPR argument.

Given the possibility of superposition, the joint state of two atoms could be (High) + (Low) for each in general, and then we can write the joint state algebraically, putting labels 1 and 2 for the particles, as

$$[(H) + (L)]_1\ [(H) + (L)]_2 = (H)_1(H)_2 + (H)_1(L)_2 + (L)_1(H)_2 + (L)_1(L)_2.$$

Each of the four terms represents a random observational possibility, with equal probability, of the states of both atoms observed: They could both be High, Low or one could be High while the other is Low. However, consider a situation where we know from physical considerations that the joint state is restricted due to a prior interaction. Perhaps one being in the state High prevents the other being in the state High. Then a valid joint QM state is $(H)_1(L)_2 + (L)_1(H)_2$. However, this cannot be written as a product of a state for atom 1 and another for state 2, side by side. There is no state whatsoever for each that can be represented in QM because $(H)_1(L)_2 + (L)_1(H)_2$ cannot be written as $[S]_1[S]_2$, where $[S]_1$ and $[S]_2$ represent *some* (any) state for particle 1 and 2. The essential point is that if particle 1 has some general state, some arbitrary superposition of (H) and (L), and similar state for particle 2, the joint state will always have all combinations, (H)(H), (H)(L), (L)(H) and (L)(L). If one of them was either (H) or (L), then also there will be terms like (H)(H) or (L)(L). There is no way one can get just the two terms (H)(L) and (L)(H) if both particles had any representable physical state in QM.

The crucial QM ingredient here is superposition, hidden within the algebraic demonstration, which prohibits a familiar model for the source as one that always gives out pairs of particles such that when one particle is in the (H) state the other is in the (L) state, each random, yet preserving the correlation. Then the only measurement combinations one can have are (H)(L) or (L)(H), randomly. Classically, such a model is reasonable, and that is precisely what was done by proponents of hidden variable theories. However, that naïve solution falls way too short in many ways.[13,14] There is a fundamental difference in specifying the physical state as $(H)_1(L)_2 + (L)_1(H)_2$, which is a superposition, and simply as a mixture of pairs of particles with the state either as (H)(L) or (L)(H). There are other measurable properties in the same pair of particles with states like (U) and (D), and clever combinations of measurements for different properties have entirely different predictions in a theory with superposition (QM) and in the naïve classical

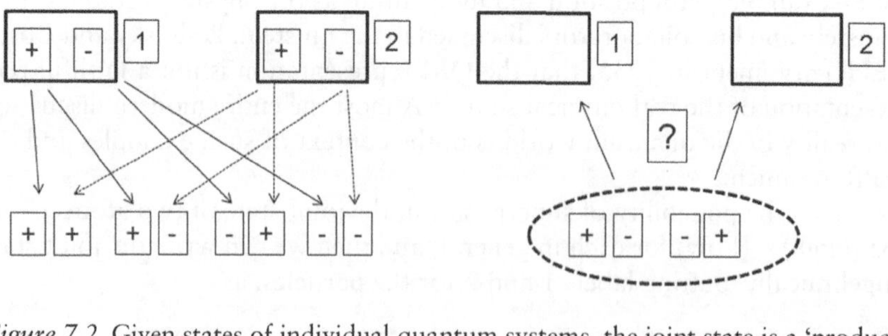

Figure 7.2 Given states of individual quantum systems, the joint state is a 'product' of both states. However, a joint state does not always factor into product states and in such a situation, there is no representation of individual states within quantum theory, in spite of the existence of a joint physical state.

hidden variable theory. That is the content of the celebrated Bell's theorem, and experiments have confirmed that the results of the measurements do not agree with what the classical statistical theory expects. The difficult and disturbing reality of QM superposition is here to stay and entirely new ways of thinking will be required to understand it, which would be a grand relief that many still hope for.

If one asks what the QM state of particle 1 is, with the joint state specified as $(H)_1(L)_2 + (L)_1(H)_2$, there is no answer, since the possible state of the particle is (H), (L) or in general, (H)+(L), and there is no way to factor out any of these from this joint state. A similar conclusion is arrived at for the joint state represented in terms of another physical property as in $(U)_1(D)_2 + (D)_1(U)_2$. Therefore, individual particles in this case have no QM state, even though the two together have a valid state. Normally, we would have expected that if there are two particles, each should be in *some* physical state, even if the details are unknown to us in some or all aspects. However, this expectation is broken in the theory of quantum mechanics. This is really surprising, since the two particles exist in the material sense, accessible for separate observations with a suitable apparatus, and the joint state exists, but each particle taken individually has no state at all in the QM representation! Such a state is called an entangled state. Quoting from Schrödinger,[15] who introduced the term, 'When two systems, of which we know the states by their respective representatives enter into temporary physical interaction due to known forces between them, and when after a time of mutual influence the systems separate again, then they can no longer be described in the same way as before, viz. by endowing each of them with a representative of its own. I would not call that one but rather the characteristic trait of quantum mechanics, the one that enforces its entire departure from classical lines of thought. By the interaction the two representatives (or [wave]-functions) have become entangled'.

What does this imply for the reality of the atom and the reality of its properties?

Here, one needs to tread very carefully, because many accomplished physicists have slipped on the trail. It is perhaps better to state what this state of affairs does not imply in the context of the nature of reality. The first thing to notice is that not being able to represent the physical state of individual particles does not imply that there is no reality to the individual particles themselves, in a material sense, and this is evident from the fact that the joint description does have components labeled by the identity of each particle. However, within such QM representation, each individual particle has no specific state, even an unknown state. Since QM is the only theory known today that accurately describes the statistical features of the microscopic world, with no known exceptions, does this mean also that there could be matter without any physical state of being, not being in any region of space, not being in some unknown direction of spin etc.? This is difficult to answer because while common sense abhors such a possibility, the only theoretical framework that we are able to work with to address such questions does explicitly point to that possibility, taken at face value.

Could this situation, then, be a limitation of our theory? Could it be that QM is not capable of describing the entirety of reality in the atomic world? Much discussion has taken place along that line as well. However, the present majority thinking is that QM mechanics is indeed the correct theory, with exhaustive scope. If that is the case, the only possibility that remains is to refine our understanding and interpretation within the theory. The other possibility, that the theory is incomplete, was argued by Einstein vigorously.

It seems that all this boils down to understanding what the wave function is in an ontological sense. It is important to note that the comparison of reality can happen at two different levels in this discussion. One is a comparison between the situation in classical physics and in quantum physics. The other is within QM itself, how the notion of reality as encoded in a representation of physical state gets blurred when one goes from single particle states to multi-particle entangled states.

How does one explore the reality in the case of two material particles, each individually addressable with experimental apparatus, but neither describable by any individual physical state whatsoever within QM before a measurement? If an observation is made on one of the particles, it does return some result from a possible set for physical quantities like energy or spin, and therefore it acquires a specific representable state within QM. Since there is a tight correlation of states, knowing the state of one implies knowing the other as well, simultaneously, without even an observation, and then a particle that had no QM state and could be far away from the region of observation also acquires a QM state! Going back to the specific example of the state $(H)_1(L)_2 + (L)_1(H)_2$, an observation on the first particle can return the value H or L randomly, but once a value is observed (say, H), then the other particle necessarily will have the complementary value (L, in this case), even if we do not make an observation. In effect, seeing H on the first particle 'collapses' the joint state to just $(H)_1(L)_2$, which is just the individual states of the two particles written side by side. Clearly, each now has acquired a specific state of its own, *from a situation of possessing no state*, in QM. The crucial point is that a material entity without any QM representation for its physical state gets a QM state as a result of a measurement on another correlated particle, possibly spatially far away. Note that instead of the physical quantity with values H or L, the observer could have chosen to measure on one particle the property that returns values U or D, and then the distant particle would have acquired either the state (U) or the state (D), depending on the outcome of the measurement. Since the physical quantity to be observed in the measurements is free choice and since the exact QM state after the observation depends on this choice, it turns out that the QM state of the distant particle is somehow determined by a free and random choice done far away from the physical system. If we assume that an instantaneous physical influence is impossible and cannot be responsible for this (assumption of Einstein locality), then we have to conclude that QM does not have a faithful and complete representation of the actual physical state of a material particle. That is, after assuming that the factual physical state of the system (if there is such thing) cannot be influenced from a distance, if the QM representation of the state is indeed influenced by a measurement performed far away (as we discussed), then there is no good correspondence between the physical state and the QM representation. This was the EPR argument of incompleteness of the QM representation.

In the evolution of the universe where every particle or constituent element shares a history with others, the entanglement and loss of individual reality in the quantum mechanical representation mean that most contents of the universe taken individually have no physical reality at all, *within QM*. An observation really does not help in improving this situation within the framework of QM because observation itself is an interaction that simply entangles the observer and the observed, dissolving the QM individuality of both! Therefore, the question of why we seem to be able to identify the reality of the individual is an unresolved puzzle with QM. This is same as the problem of emergence of the apparent classical world from the underlying, more fundamental, quantum world.

One may say that the assertion (or belief) that there is some physical state for every physical system, however unknown the details may be, can of course be doubted and questioned. In fact, if this belief is not denied, then QM is already proved to be incomplete by the EPR argument! This is precisely where all the efforts to understand the nature of reality within QM are jammed. However, the majority of physicists go further and believe that QM representation is same as the actual physical state, QM being the complete physics of the situation. Therefore, they believe that the state of affairs one finds in the case of entangled systems implies that there are instantaneous influences (nonlocality) over spatially separated regions that change physical states and their QM representation. This is a serious situation because there is no empirical evidence at all that there is any such a nonlocal influence and the belief is based on trying to translate what happens to QM representations into concepts familiar from classical statistical theories in real space and time. I may also add that a severe misunderstanding of what the Einstein (EPR) argument and the remedy he hoped for have led to what are called local hidden variable theories. These theories break explicitly the basic QM features like superposition, and yet hope, naïvely and ignoring the empirical and logical reasons for QM, to describe the physics of the microscopic world in classical statistical terms. Physicists went to great lengths of experimental effort to rule them out.[16] Such are the unfortunate consequences of having to use an unobservable in QM, whose ontology remains by and large ill understood.

From unobservables to unspeakables: EPR and physical reality

A brief discussion clarifying the concept of physical reality in the context of the EPR assertion of the incompleteness of quantum mechanics is perhaps appropriate. My motivation for this discussion is the enormous volume of literature on this subject containing painfully severe misunderstandings of the EPR argument and then equally disturbing assertions about what it implies for reality and locality, the fundamental pillars of classical physics. The misunderstanding can perhaps be ascribed to the historical accident that the published paper in the American journal *Physical Review* (ref. 12), from which people get these ideas, was written by Podolsky (for reasons of language), and it does not faithfully represent Einstein's essential argument because his 'main point was, so to speak, buried in erudition'.[17]

Indeed, the argument is crystal clear when presented in Einstein's own words. The articles 'Physics and reality' (ref. 3) in the *Journal of the Franklin Institute* (1936) and 'Quantum mechanics and reality',[18,19] written in 1948 for the journal *Dialectica*, are authentic sources (similar to that in his letters to Schrödinger and Popper in 1935, subtle differences notwithstanding). Einstein considers first a wave function for one particle that does not specify a sharp position or momentum and asks whether the particle really has a sharp position and momentum 'in reality' but the wave function does not behold that reality, or is it that the entire reality is the un-sharp specification that the wave function represents. The former case implies that the representation is incomplete and also that uncertainty principle does not hold for the reality whereas the wave function description obeys it. In the latter case, which physicists then and now subscribe to, the realisation of a sharp position on observation can be then attributed to the measurement process. Einstein admits that this view alone 'does justice in a natural way to the empirical state of affairs expressed in Heisenberg's [uncertainty] principle within the framework of quantum mechanics'. Then he makes the crucial observation that in this standard view, two wave functions that differ in more than trivialities describe two different real situations. It is important to notice that Einstein did not bother to define what reality is etc., as done in detail the EPR paper, 'burying the main point in erudition'. The next point is the 'separation principle', which is also the basis of the principle of Einstein locality. It is the idea of the independence of the existence of objects that are far apart from one another in space. Then external influence on one at location A, like the action of a measurement, has no influence on the other at location B. Then he notices that the QM description of the two-part physical system is in general in terms of the joint wave function ψ_{12}, which cannot be written in terms of wave functions for the two independent systems $\psi_{12} \neq \psi_1\psi_2$ and points out that the wave functions for the single-part systems do not exist at all. The methods of quantum mechanics, however, allow the determination of ψ_2 by making a suitable observation on system 1 (S1) without a measurement on system 2 (S2), determining both ψ_1 and ψ_2. However, the nature of the resulting ψ_2, whether it corresponds to a sharp position or a sharp momentum, for example, depends on what measurement is carried out on system 1. Hence, he concludes, that according to the choice of measurement on S1, a different real situation is created in regard to S2, because different wave functions correspond to different realities. Because we have assumed already that physical reality cannot be altered or created by an act of observation that is spatially separated, the ψ functions cannot be a faithful and complete representation of physical reality.

That completes Einstein's version of the EPR argument. A reading of the EPR *Physical Review* paper immediately shows why it does not faithfully represent Einstein's argument and why it is misleading and unnecessarily erudite.

From wave functions to quantum fields: vacuum vs. cosmos

Quantum theory of physical systems consisting of large number of particles led to the concept of a quantum field associated with the particles, a generalisation of the wave function itself. These fields, one kind for each particle we know of, with the QM version of the electromagnetic field serving as a prototype, are fundamental and essential to modern physics, with many success stories of precision calculations. Also, they are considered unavoidable if the quantum theory has to be formulated consistently with the theory of relativity. These fields are as unobservable as any other fields we considered, even though their space-time status is better than that of the wave function familiar from quantum mechanics of a few particles. One of the fundamental features of such a field theory is that there is an infinite amount of 'zero-point' energy in the vacuum state of these fields. The vacuum state is a situation where there are no real particles apparent. Since these fields are modelled after oscillators with all possible frequencies and since an oscillator in QM has a zero-point energy (energy in the lowest possible state) worth $hf/2$, where h is the Planck's constant and f the frequency, an infinite or even a large number of oscillators in the field has a very large, near-infinite amount of energy. This cannot be avoided. The early impression was that this really does not matter, since constant background energy has no observable consequence. The general attitude was that what cannot be observed, even an infinite amount of energy, need not bother us, even though all this came about because we had to use some unobservables in the theory.

However, this relief was short lived because it was realised that there is in fact one dramatic observable consequence of a constant background energy. The rate of expansion of the universe, or the rate at which galaxies move away from each other, is directly proportional to the average energy density in the universe, and nothing can be added or subtracted without affecting this observable and measured rate. This expectation from theory (the general theory of relativity) is verified at least approximately even if we do not make new hypothesis about dark matter and so on. Even though the amount of matter that is visible through the light emitted is estimated to be only 5% of what is required to explain the observed rate of expansion, we may say that there is an approximate consistency. In contrast, the vacuum zero-point energy calculated in the context of quantum field theories far exceeds, by numbers beyond imagination, what is reasonably consistent with the measured expansion rate of the universe. Thus, cosmology provides a powerful test of our theories, and the verdict goes against the theoretical structure (of quantum fields) we use today. No solution has been found, and a solution may well have to deny the reality of quantum fields. This will also mean that we will have to reconstruct the relevant physical theories in terms of quantum fluctuations and energy of only matter, without the zero-point energy of the quantum fields. At present, it is not known whether this is possible even in principle.

Some aspects of the 'successful' standard model of particle physics, which depends on what is called 'gauge freedom', or freedom for changing the values of certain field quantities in the theory without observable effects, also may have to be reconsidered in detail in the context of the consequences for the dynamics of the universe.

Postulates as reality

Postulates in physical theories are formal general statements on fundamental physical phenomena, usually and preferably extracted from observational evidence. What is peculiar in physics is that a lack of falsifying evidence also can be taken as supporting evidence for a postulate based on which the theory is constructed. If the theory turns out to be consistent with observations, especially those that come after the construction of the theory, the faith in the postulate is strengthened. Since accessible and realisable observations and experiments are vastly limited in comparison to the scope of a general theory, this situation allows postulates that are familiar and not easily falsifiable to be projected as truth and reality. The ill effects of this scenario are more and more evident in present-day physics, with its many postulates and several unobservable entities outside the immediate reach of experimentation.

What will be surprising is that even the most familiar and all-pervading fundamental physical theories—the special theory of relativity—suffers from this problem. Its fundamental postulate of the absolute constancy of the velocity of light relative to inertial observers is neither proved nor contested, but most physicists believe that it is amply verified to be true. The real nature of the physics of light and relativity is intimately related to the universe and matter and therefore, we turn to a discussion of a new and essential paradigm of physics in relation to the universe in which it is formulated, tested and applied.

Universe, space, time and notions of reality

All our fundamental theories, without exception, were constructed and completed before we acquired any significant knowledge about the universe—its matter content, extent, history and dynamics. There are serious philosophical and physical issues involved here because all our theories assume, and in some cases require, empty space as their backdrop. However, the matter-filled universe is also very gravitational, and since we know now that gravitational interaction can change the rate of clocks and the length standards, the entire edifice of metrical physics needs a serious reconsideration.

To start with, all our theories are conceived, completed and tested in the presence of all the matter in the universe, and if the relevant gravitational effects are not included and accounted for systematically, we could be working with deficient or incomplete theories, even when there is apparent agreement between what we expect from theory and what we see observationally. All experimental results naturally include any cosmic gravitational effect that may be there, since all experiments are done in the unavoidable presence of such matter, whereas the theories presently used to interpret them assume an empty space-time as their background. A most plausible scenario is that many physical effects we see are in fact gravitational effects linked to the matter in the universe, whereas our theories describe them as being due to other reasons. This indeed is the case that is now evident from several considerations, but we will discuss only one or two relevant points here, since our focus in this section is on examining the status of space as a physical entity in physical theories. The full programme to formulate physics in the 'once given universe', called Cosmic Relativity, has many interesting features and predictions for relativity effects, electrodynamics and propagation of light.[20,21,22]

Physicists seem very proud of their achievement of banishing the old 'ether' from physics, and this is usually highlighted as an example of removing unobservables from physical theory. In fact, there are really no good empirical reasons that are brought forth in support of this achievement, and the typical reasoning quotes the success of special theory relativity that rendered the ether irrelevant. Treated rigorously from an empirical point of view, an unobservable like ether could never be 'disproved'. Often-quoted remarks that the null result of the Michelson-Morley (M-M) experiment and its variants disprove the reality of ether are based on superficial understanding of both the physical analysis of the experimental result and the history of the issues involved. In fact, what happened was what Poincare predicted. In *Science and Hypothesis*, he wrote that[23] 'whether the ether exists or not matters little—let us leave that to the metaphysicians; what is essential for us is, that everything happens as if it existed, and that this hypothesis is found to be suitable for the explanation of phenomena. After all, have we any other reason for believing in the existence of material objects? That, too, is only a convenient hypothesis; only, it will never cease to be so, while some day, no doubt, the ether will be thrown aside as useless'. Special relativity, however, certainly deviated from the premise that 'everything happens as if the ether existed'.[24] Einstein's hypothesis of constant relative velocity for light is an elegant solution of the null result of the M-M experiment, and the Lorentz-Fitzgerald contraction hypothesis is another valid solution. Ironically, the experimenters continued to write that they were 'proving' the Lorentz hypothesis through their null results well into the 1930s. Since they were in the very laboratory frame in which the experiments were performed, special relativity denies any length contraction or time dilation effects, whereas all experimenters preferred to depend on such physical effects to explain their null results.

The success of the special theory of relativity brought empty space and the vacuum as major physical entities into physics. Not realising that there was no empty space in reality and that the entire universe was filled with matter in some form allowed a misguided enthusiasm about bestowing reality status to empty space and an associated time, together called space-time, even though there is no way to think about these without matter. Gravity was interpreted as the geometrical distortions that happen to space-time due the presence of matter or by a prior design. No doubt, the result was an elegant theory with great success, but it alienated the physics of gravity from the physics of other interactions that could not be formulated the same way. Fundamental to these developments was the equivalence principle, which is the universal relation between the 'charge' of the gravitational interaction and 'inertia' in dynamics, and both were identified and called by the same name, 'mass'. In this course, Ernest Mach's brilliant insights (ref. 7) on the possible origin of inertia as due to interaction with the matter around was dismissed as irrelevant, in spite of a hint of evidence in favour of general relativistic effects related to rotation.[25] Later, this was distorted to mean that what Mach said was incorrect. Throwing out the unobservable ether is one thing, but daring to throw out the readily observable matter-filled Universe from consideration is a colossal error of judgment. In any case, the general theory of relativity is considered our best theory, and it is well tested. However, the aura about its geometrical nature with stress on empty space and time as physical entities is in fact overstated. This could very well be the red herring that delays the goal of a quantum theory of gravity.[26] It is even likely that the geometrical interpretation of its application to the dynamics of the universe could contain some features that will need revision in the future.

Special relativity: the theory built on several unobservable pillars

It is interesting to note that even real physical length contraction and time dilation are unobservables for the observer moving with the scale and clock because all such physical measurements are *comparisons* and the standard scale and clock used for the comparison also suffer exactly the same contraction and dilation. This unobservability for a comoving observer is of course different from the universal unobservability of some theoretical entities ('free creations of the human mind') we discussed. It is precisely the unobservability of these real physical effects that allows the special relativistic assertion that there is in fact no contraction or dilation in the comoving frame. Once the contraction is denied, then and only then can the invariance of the relative speed of light be postulated as the solution.

This discussion shows how easily one can be misled about the reality and truth of phenomena in physical nature because of our equating reality to what could be observed. Realising that a measurement is often a comparison with a physical standard that is subject to the same physical effects we are trying to measure is of the utmost importance in evaluating and testing a theory with empirical and logical rigour. Physicists who claim that standard tests of the M-M type prove special relativity and the constancy of the relative velocity of light lack this rigour. In fact, as soon as *round-trip comparisons* are included as possible tests of relativity theories, in which case cumulative effects like time dilation become observable, the weakness of special relativity is revealed.[19,27,28] Clearly, 'unobservable' in physics does not always imply 'unreal'. Reality cannot be denied just because the phenomena are not observable by a limited and constrained class of observers.

The special theory of relativity implements the principle of relativity—the invariance of the laws of physics in all inertial frames—by postulating the absolute invariance of the relative speed of light in all frames in empty space. Its supporting pillars are two unobservables: empty space (and its velocity independent isotropy) and the absolute invariance of the relative one-way speed of light. Einstein realised that the one-way speed of light over a stretch of spatial distance, which requires two clocks, is an unobservable in a limited sense. Since this measurement requires synchronisation, the one-way speed depends on the convention for synchronisation. Though only the two-way speed of light is experimentally shown to be an invariant, it is possible to postulate that even the one-way speed is an invariant by suitably defining a synchronisation embodied in the first-order term of the Lorentz transformation. In fact, if the time light takes to cover a distance Δx is $t = \Delta x/c$ in a stationary frame, it should take an *additional* duration

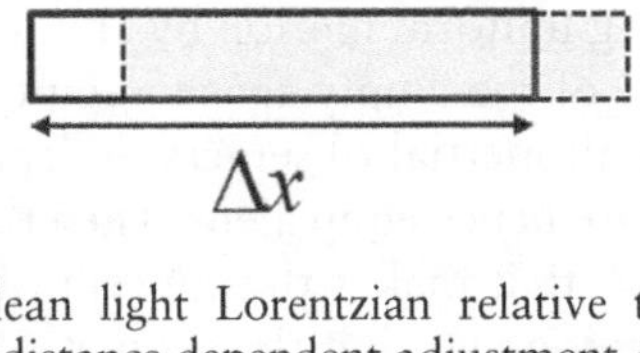

Figure 7.3 Making Galilean light Lorentzian relative to a moving frame by a direction and distance dependent adjustment $\delta t = v\Delta x/c^2$ of clocks.

of $v\Delta t/c = v\Delta x/c^2$ relative to an observer moving at velocity v *if the relative velocity of light is not a constant.*

However, if we postulate that the clock at the distance Δx requires a distance-dependent adjustment for some reason and decree that it is to be adjusted back by the amount $\delta t = v\Delta x/c^2$, the additional duration that (Galilean) light takes is cancelled and the apparent duration is just $t = \Delta x /c$, which is the same as the duration relative to an observer at rest (Figure 7.3). This is exactly like the Galilean adjustment of the moving coordinate reference x' where the coordinate values are compensated for the movement at velocity v for duration t as $x' = x — vt$. But this adjustment of the clock allows us to claim that the speed of light measured with such a (fake) clock remains invariant to first order in v/c! Herein lies the circularity of the reality created by the special theory of relativity: The propagation of light is used to define the time of separated clocks, and the same time is used to claim the invariance of the relative speed of light. (It is also relevant to note that the first-order 'correction factor' is exactly what an absolute frame theory would give from *its second-order effect* for the time shown by a clock synchronised at one end of the scale Δx and then slowly moved to the other end, when the entire system is moving at speed v.) It is important to stress the factual situation that no experiment has ever shown, independent of conventions or circular arguments, that the one-way speed of light is an invariant constant. (The smaller second order effect, $(1\text{-}v^2/c^2)^{1/2}$, is in fact the only empirically verified correction to moving clocks after round trips.) In fact, the moment we can identify a universally synchronised clock—any phenomenon that serves as a time-keeping device, which is the same everywhere in an extended spatial region—the hypothesis of the invariance of the one-way speed of light can be tested. Such a 'Galilean clock' will also be in conflict with the Lorentz transformation because it will show that the clock adjustment $\delta t = v\, \Delta x /c^2$ at distance Δx is fake. This identifies exactly where the weak point of the theory lies.

Relativity and the universe

A serious conflict arises between what we know about the universe today and what was constructed as the 'correct' theory of relativity in 1905. The special theory of relativity described all relativistic effects as due to relative motion between different inertial observers and maintained that the notion of special preferred frame, or an absolute frame of reference, cannot be sustained. In the special theory, the validity of the principle of relativity (the impossibility of detecting uniform motion by any physical phenomena) was linked to the invariance of the empty space—empty space remains homogenous and isotropic to all inertial observers, independent of their velocity, and all these observers are hence equivalent. Then there is no universal time, nor a universal reference that makes the concept of absolute rest meaningful. Indeed, just the requirement that the geometry (metric) of empty space remains invariant under motion leads to the Lorentz transformations that form the entire basis of the theory. However, real space as we observationally know it today is not empty at all. It is filled with matter. Therefore, the space appears very different to a moving observer in comparison to an observer who is stationary relative to the average matter distribution. In relation to a moving observer, the entire matter moves as a large directional current and the space appears anisotropic, proportional to the velocity. This is in fact easily observable by a moving observer as the 'dipole anisotropy' of the cosmic microwave background radiation. There is indeed operational and conceptual meaning to a state of absolute rest in such a situation. Since space does not remain isotropic in a moving frame, the underlying geometry is clearly anisotropic in moving frames and the inconsistency of the Lorentz transformations to describe the situation is immediately evident: Lorentz transformations preserve isotropy and homogeneity, whereas isotropy is broken under motion in real matter-filled space. There is even a universal absolute time, since the evolving matter and radiation that define time have essentially the same history everywhere in the universe. Operationally, the slowly decreasing temperature of the cosmic microwave background radiation can be taken as this universal time, and it is automatically synchronised everywhere in the universe. Thus, we have identified a reasonably precise and useful Galilean clock (ref.22,28). Indeed, the whole premise of special relativity is made invalid by the evolving observable matter-filled universe. Replacing the unobservable ether and the even more unobservable empty space with an observable matter-filled universe as the background arena of physics restores logical and empirical consistency of physical theory with the gravitational presence of the universe.

One can go further and answer several questions that arise. The anisotropic space and its geometry in the frame of the moving observer can be correctly described by the Galilean transformation and, along with the gravitational action of the matter and matter current in the universe, all observed relativistic effects like time dilation are explained (ref.22). This framework also shows that inertia is indeed the resistance to acceleration arising in the gravitational interaction with the entire matter in the universe, as Mach had guessed. Most surprisingly, Newton's law of motion emerges naturally as a gravitational effect on motion.[29] Along the way, the principle of relativity and the equivalence principle, hitherto considered as essential postulates for physics in the unobservable empty space, *materialise* as consequences of the gravitational action of the matter-filled universe. The lesson from this development is that relativity without cosmic matter is empty relativity with no physical effects on moving clocks etc. Physics in empty space is meaningless. Matter and its gravity indeed are the real basis for relativity and dynamics. With this realisation, and several empirical consequences and supporting evidences in the context of detailed behaviour of clocks and the one-way propagation of light, the concepts of absolute rest, absolute space and time are firmly back into physics, and the issues of reality are linked inseparably to the reality of matter and its gravitational action. The most drastic consequence of the return of the absolute is understandably in the behaviour of light. Unambiguous empirical evidence for the first-order non-invariance of the relative velocity of light (ref. 21,22), while preserving the upper limit of c for all motion in the cosmic frame, restores the underlying reality of a cosmic matter-based paradigm of relativity and dynamics. That we continue to use the unobservable gravitational field for a convenient description of this scenario is another matter!

Concluding comments

It seems that the use of unobservable entities is unavoidable for theorising about physical nature. However, it is very important to be watchful and critical about their overall consistency within the structure of physics as well as their implications to all conceivable physical situations. The problem is not really the use of unobservables, but the insistence on non-critically extrapolating from the resulting successes of the physical theory on a limited set of predictions to the unquestionable reality of the unobservable. Unobservable wave functions and quantum fields in modern physics emerged naturally from theoretical needs, but their ontological status is a source of constant debate and worry. While the wave function can be lived with, but for the bitter feeling and frustration of not being satisfied with sufficient understanding of the space-time picture of cause-effect relations in the microscopic world, the damaging inconsistency of some features of the quantum fields with the dynamics of the universe is something that needs curing in a future theory. It is clear that the inseparable and ever-present gravitational link to the matter-filled universe is in conflict with a theoretical framework that relies on the unobservable empty space and the associated time as its basis of physical effects. In fact, physical changes that have no cause in physical interactions are suspect, and they always point to the need for a new theory. At least in this case, a much better realistic basis is provided by the paradigm of Cosmic Relativity that derives all relativistic physical effects as due to matter and its gravity in the universe. With this realisation, the absolute frames of space and time are back in physics. Not surprisingly, this paradigm with its universal cosmic links and physical inseparability also holds human concepts of harmony and beauty.[30]

While a programme to banish all unobservables from physical theory is unlikely to succeed, and perhaps even unnecessary, they are both the strong and weak points of theoretical descriptions of physical nature. Analysing them for empirical and philosophical consistency and integrity is always a promising path towards a better theory.

Acknowledgments: I thank Partha Ghose and Peter deSouza for the invitation as well as for the friendly and warm hospitality at the Indian Institute of Advanced Studies, Shimla, during the superbly nourishing seminar, 'The Nature of Reality: The Perennial Debate'. Discussions with several participants, especially the philosophers who expect better rigour, have spurred and helped me to sharpen some of the thoughts presented in this chapter. Conversations with Martine Armand on the contents and style have helped in a clearer presentation.

Notes

1 Unnikrishnan, C. S., 'Universe from beyond: Role of unobservables in physics', in *Science and Beyond: Cosmology, Consciousness and Technology in Indic Traditions*, S. Menon, B. V. Sreekantan, Anindya Sinha, Philip Clayton and R. Narasimha (eds.) (Bangalore, India: National Institute of Advanced Studies, 2004), pp. 224–29.
2 Unnikrishnan, C. S., 'The role of unobservables in modern physics and our links to the physical universe', in *Traditions of Science: Cross Cultural Perspectives*, P. Bilimoria and M. K. Sridhar (eds.) (New Delhi: Munshiram Manoharlal Publishers Pvt. Ltd., 2007), pp. 317–23.
3 Einstein, A., 'Physics and reality', *The Journal of the Franklin Institute* 221 (1936), pp. 349–82.
4 Einstein, A., 'The problem of space, ether, and the field in physics', in *Mein Weltbild (My Worldview)* (Amsterdam: Querido Verlag, 1934).
5 Jammer, M., *Concepts of Space: The History of Theories of Space in Physics* (Cambridge, MA: Harvard University Press, 1954).
6 The first scholium after definitions in Newton's *Principia* (Book 1) attempts to grasp the primary concepts of space and time.
7 Mach, E., *The Science of Mechanics* (London: The Open Court Publishing Co., 1942).
8 d'Espagnat, B., *Reality and the Physicist* (Cambridge: Cambridge University Press, 1989).
9 Pais, A., 'Einstein and the quantum theory', *Reviews of Modern Physics* 51 (1979), pp. 863–914.
10 This refers to a famous query by Einstein on reality in the QM description, quoted in reference 8.
11 Ghose, P., *Testing Quantum Mechanics on New Ground* (Cambridge: Cambridge University Press, 2006).
12 Einstein, A., Podolsky, B. and Rosen, N., 'Can quantum-mechanical description of physical reality be considered complete?', *Physical Review* 47 (1935), pp. 777–80.
13 Bell, J. S., *Speakable and Unspeakable in Quantum Mechanics* (Cambridge: Cambridge University Press, 1988).
14 Unnikrihnan, C. S., 'Conservation laws, correlations functions and Bell's inequalities', *Europhysics Letters* 69 (2005), pp. 489–95.
15 Schrödinger, E., 'Discussion of probability relations between separated systems', *Proceedings of the Cambridge Philosophical Society* 31 (1935), pp. 555–63.
16 Afriat, A. and Selleri, F., *The Einstein, Podolsky and Rosen Paradox in Atomic, Nuclear, and Particle Physics* (New York: Plenum Press, 1999).
17 Einstein, in a letter to Schrödinger (1935).
18 Einstein, A., 'Quantum mechanics and reality', *Dialectica* 2 (1948) pp. 320–24.
19 Born, I., *The Born-Einstein Letters: Correspondence Between Albert Einstein and Max and Hedwig Born from 1916 to 1955* (London and Basingstokes: Macmillan Press Ltd 1971).
20 Unnikishnan, C. S., 'Cosmic relativity: The fundamental theory of relativity, its implications, and experimental tests': http:arXiv:gr-qc/0406023 (2004).

21 Unnikishnan, C. S., 'Moving clocks, twin clock problem and the universe', in *Physical Interpretations of Relativity Theory: Proceedings, London 2006*, M. C. Duffy, V. O. Gladyshev, A. Morozov and P. Rowlands (eds.) (Bauman: Moscow State Technical University, 2011), pp. 493–504.
22 Unnikrishnan, C. S., 'Physics in the "once-given" universe', in *Recent Developments in Theoretical Physics*, S. Ghosh and G. Kar (eds.) (Singapore: World Scientific, 2010), pp. 99–120.
23 H. Poincare, *Science and Hypothesis* (New York: The Walter Scott Publishing Co. Ltd., 1905).
24 H. Brown, *Physical Relativity* (Oxford: Oxford University Press, 2005).
25 Unnikrishnan, C. S., 'Experiments motivated by the Mach's Principle: A review with comments', in *Mach's Principle and Inertia*, M. Sachs and A. R. Roy (eds.) (Montreal: Apeiron, C. Roy Keys Inc., 2003), pp. 89–98.
26 Unnikrishnan, C. S., 'Quantum gravity and the structure of physical reality', in *History of Science and Philosophy of Science (PHISPC Volume XIII Part 6)*, P. K. Sengupta (ed.) (New Delhi: Pearson-Longman, 2010), pp. 409–44.
27 Sagnac, G., 'The demonstration of the luminiferous aether by an interferometer in uniform rotation', *Comptes Rendus* 157 (1913), pp. 708–10, 1410–13.
28 Unnikrishnan, C. S., 'On Einstein's resolution of the twin clock paradox', *Current Science* 79 (2005), pp. 2009–15.
29 Unnikrishnan, C. S., 'True dynamical tests of the equivalence principle', *International Journal of Modern Physics: Conference Series* 30 (2014), p. 1460267 (9 pages).
30 Unnikrishnan, C. S., 'Contemplations on our physical links to the universe: Searching for and finding the hidden harmony', *Leonardo* 39 (2006), pp. 71–5.

8 High-energy physics and post-empiricism

K. Sridhar

Once we examine the simplest elements, where would be the boundary between science and philosophy at which we could stop?

Ludwig Boltzmann[1]

Preamble

It is remarkable that almost a century has elapsed since the time Tagore and Einstein had what is now a historic dialogue, but the deeper issues that concern us even today in the sciences, social sciences and the humanities are still very much the same as those which the two great men discussed. These were questions primarily about truth and beauty, and they approached it from opposite perspectives: the oriental and the occidental, the arts and the sciences. Nevertheless, what engaged both these thinkers was the human question. For Tagore, the sciences and the arts were part of human enterprise and the notions of truth and beauty were contingent and did not have a meaning independent of the human agent, but while Einstein conceded that beauty may be contingent, he argued that truth endured beyond any human condition. The truth of mathematical and scientific statements was not limited by human subjectivity.

While, in science, it is Einstein's view of an objective truth that has dominated it, is probably important to pause and pay attention to Tagore's appeal to see the enterprise of truth-seeking in science as a very human enterprise, especially because, unlike mathematical truths, scientific truths are empirical and require validation in experiments. Mathematical truths are analytic and *a priori*, whereas the synthetic truths of science can be validated only *a posteriori*. Can truths of science be valid *a priori* without a need for experimental justification? This last question has been around at least since Kant, but it seems that it has become a major concern with physicists and philosophers of physics in recent years. It is this question that we will take up in the present chapter.

Introduction

I start this chapter with a somewhat provocative question: Is there practice in the method of science? This is a question not about whether there is method in the practice of science, but about whether questions of science methodology pay sufficient attention to the practice of science. Philosophical and methodological issues rarely interest scientists; these are the concerns largely of philosophers and social scientists. So when scientists are drawn into a discussion of these issues, they often enter these debates invoking fixed and old-fashioned notions of these issues. Any new perspective or formulation is met with much suspicion by scientists, but very often, they have nothing to offer as an alternative either. On the other side, philosophers and social scientists often fall back on dated scientific theories and fail to keep up with emerging paradigms. Questions of science, its nature and its method cannot be frozen in a classical eighteenth-century understanding given the way science itself has changed over the ages. Formulations of the philosophy and methodology of science need to engage with science as it is practised, and this can happen only if there is a two-way exchange between philosophers and practitioners of science and if a shared language within which to discuss these ideas is developed. What exists now is a seemingly insurmountable gap that was developed and cemented through the 20th century, when the Anglo-American traditions of science have dominated. The Boltzmann quote cited in the beginning of this chapter speaks of another time and another tradition. That was a statement he made at a time when atomism had opened up a foundational crisis in the subject, but the integrated approach adopted at that time by physicists, mathematicians and philosophers alike had much bearing on how the story of atomism eventually panned out.[2] I will demonstrate in this chapter that a similar impasse has again been encountered recently in areas of physics like high-energy physics and cosmology; in particular, and the separation of knowledge systems that has been systematically achieved in the last century deepens the crisis.

Going post-Popper

In the last year or so, a fresh and vigorous debate about the epistemic nature of science has erupted in journals and on discussion pages on the web, and several physicists have taken part in this exchange. To begin with, this is already a healthy indication and possibly a harbinger of many such discussions around the epistemology and methodology of science. But from some of these articles and also from the reactions that other physicists have posted in response to the web articles, the absence of a deeper understanding becomes all too obvious. Hopefully, this debate will spur other equally important ones and will allow for the development of a scientific discourse informed by philosophy and other human sciences.

The principal protagonists on one side of the debate have been the philosopher Richard Dawid and the physicist Sean Carroll. Dawid has been proposing a new, post-empirical approach to string theory. He has developed this position in a series of papers[3] and in a recent book.[4] I will discuss his position in more detail later, but his main point is that string theory is particularly suited to non-empirical theory assessment but also mentions inflationary cosmology in this context. Dawid's argument rests primarily on what he calls 'scientific underdetermination', which is somewhat different from the idea of underdetermination that philosophers of science like Quine have proposed. For the moment, we will understand underdetermination to mean the situation where available empirical evidence is insufficient to select out one theory from a set of theories and discuss this at greater length later. Dawid argues that in theoretical enterprises like string theory, which lack any empirical support at the given moment, the belief in their validity stems from the fact that underdetermination is severely constrained.

With probably some awareness of Dawid's position but largely independent of it, Sean Carroll launched an attack on Popperian falsifiability.[5] He claims that falsifiability is an idea that has outlived its utility and has to be discarded in discussing modern science. String theory and, in general, any approach to quantum gravity, can be tested only at energy scales which are so high that they are completely out of the realm of experimental testability. The idea of a multiverse or the many-world interpretation of quantum mechanics also cannot be empirically tested in any conceivable way. Carroll is weary of a section of the physics community that deems such theories non-scientific, basing their argument on Popper and falsifiability. He believes that the hallmark of a scientific theory is two-fold: It should be definite and it should be empirical. By definite, Carroll means that the theory should say something unambiguous about the 'reality' it deals with. By empirical, he does not mean a theory should be testable and falsifiable, but rather that there should be a connection with experimental data, however indirect that may be. In his reckoning, strings and multiverses do satisfy these criteria and therefore simply invoking Popperian falsifiability to dub these theories as non-scientific is wrong. Carroll is not unaware of the demarcation problem, i.e., the criteria we need to have to demarcate science from non-science and the problem of addressing the demarcation issue if we give up on falsifiability, and he does not really suggest a way of addressing this problem. He thinks that eventual acceptance or rejection of these theories will be on the basis of whether they work effectively as theories and offer a coherent description of the physical reality they deal with.

Carroll's paper provoked considerable response in the physics community. The most prominent amongst these was a comment published in *Nature* by the cosmologists George Ellis and Joe Silk.[6] Ellis and Silk take the more conservative viewpoint in arguing that Popperian falsifiability is the hallmark of physical science and that there is no way that can be sacrificed at the altar of theoretical speculation. They argue against the claim of uniqueness of theories of strings or multiverse and cannot see any reason in according these theories any special status and subjecting them to non-empirical assessment. They do conclude by saying that they are happy that Carroll and Dawid have opened up this debate about empiricism and express the hope that the debate will be taken forward and a new 'narrative for the scientific method that can deal with the scope of modern physics' will be developed. However, they add that whatever the new formulation, it should still conform to the criteria of observability and falsifiability.

The physicist Sabine Hossenfelder has a response to this debate similar to that of Ellis and Silk, where she dubs post-empirical science an oxymoron.[7] Her response is more explicit about the politics of this debate, which she sees as being related to job appointments, tenures and funding. It is certainly a question to ask whether the entire post-empiricist defence is being put up only to allow some speculative theories to continue their dominance over the physics community.[8]

A more nuanced assessment of Sean Carroll's position comes from the philosopher of science Massimmo Pigliucci,[9] who is with Carroll to the extent that he is also against the very naïve use of falsifiability that some physicists invoke against string theory or the multiverse. He points out, citing Duhem and Quine, that falsifiability is no longer as simple as it was when originally proposed. But he thinks that if a theory like string theory has been around for a long time and still not shown any empirical yield, then the physics community is justified in rejecting these theories.

Having summarised the main points of view that have emerged in this debate, I will now go on to discuss these in some more detail.

Demarcation, falsifiability and falsification

Popper is himself very clear about the meaning of falsifiability.[10] In order to avoid the problem of induction that Hume alerted us to,[11] Popper gives up an inductivist account of knowledge. In this he breaks away from the logical positivists too, for, unlike them, he is not concerned with eliminating metaphysics altogether, but rather, he is seeking a criterion to distinguish metaphysics from scientific knowledge. Thus, he arrives at the problem of demarcation, which is that of finding a criterion to distinguish statements of science from statements belonging to what we may call 'non-science'. In the latter set of statements, Popper includes non-scientific (statements that are not science and not purporting to be science), pseudo-scientific (statements that are not science yet make claims to be statements of science), metaphysical statements and statements of mathematics and logic. Popper then introduces the idea of a basic statement,[12] which describes a possible event that may be observed. Any statement (or a set of consistent statements, i.e. a theory) is called falsifiable if one can find a possible basic statement that contradicts it. The basic statement is then called a potential falsifier. It is significant that Popper uses falsifiability rather than verifiability as the criterion for demarcation. For him, inductive inference, which is the movement from the validity of singular to those of universal statements, is not admissible. So, unlike the positivists, he does not see inductive verification as a possibility and believes that the positivist insistence on verifiability is erroneous. He consequently brings in falsifiability as a demarcating criterion.

Falsifiability, so defined in the context of the demarcation problem, is about the logical relation between a set of theory statements and a set of potential falsifiers. As a criterion of demarcation, to establish the empirical-scientific character of a theory requires one to describe a non-zero set of potential falsifiers. It becomes clear, then, that falsifiability is a logical relation between two sets of statements and has to do with the logical structure of classes of statements.

Quite different from this, though somewhat related to this, is the idea that Popper sometimes refers to as falsification, which has to do with proving an empirical scientific theory false and which is distinct from falsifiability. Popper is, of course, fully aware of the problem with falsification and how it is never possible in practice to prove a scientific theory false: 'It is always possible to find some way of evading falsification, for example by introducing ad hoc an auxiliary hypothesis . . .' and that '[n]o conclusive disproof of a theory can ever be produced'.[13] Falsifiability is an 'in-principle' logical possibility, whereas falsification is about a conclusive experimental proof to show a theory to be false. In these debates, much confusion abounds in the use of these two concepts, falsifiability and falsification, and it is best to clarify the meaning Popper attributed to these.

Finally, Popper is clear about the fact that the demarcation between science and metaphysics that he is seeking is only being done to arrive at a theory of knowledge. As mentioned earlier, he does not seek to eliminate metaphysics, but even accepts that by taking a historical view of the development of science, it is seen that some metaphysical ideas have even led to the advance of science. Thus, on the question of metaphysics, Popper differs significantly from the logical positivists.

Realism/anti-realism

Another issue that needs to be clarified is that of realism/anti-realism, and this is more complex because it is not restricted to the philosophy of science, but spans the entire spectrum of philosophy, from epistemology, causality, ethics, aesthetics, sociology, psychology and even the domain of our everyday experiences. The discussion here will be confined to the philosophy of science and to some extent the issues from the theory of knowledge that make contact with science.[14] The basic question that one starts from here is: Are the entities of science real? A related question that arises is whether that reality is independent of us human observers. The reality question has more charge, of course, with entities that are not observable. The realist's answer to the reality question is a resounding 'yes!' and her argument is that if there is a chain of links that connects any entity that science posits to a set of empirical phenomena that can be verified in experiment, then that gives us the belief in the existence of these entities. They are not explanatory devices, but real things that exist; rather, the fact that these entities have to be invoked for explaining phenomena is to be taken as a guarantee of their existence. There are various strains of realism, in fact, and they occupy a spectrum of positions. There is ontological realism, on the one hand, which claims the reality of every entity posited in a literal manner, and there are ones like structural realism, which say that if not the unobservable entities, at least the theoretical structures they give rise to have epistemic content. The realist's position can be summarised by saying that she accepts the scientific theory literally *and* she believes in the truth of this theory. Giving up either the literal construal or the truth of scientific theories is the movement towards an anti-realist position.

Anti-realists refuse to equate the validity of a scientific theory with the existence of the entities it postulates. Again, there are several different anti-realist positions, but largely they say that the entities that are posited are shorthand devices that we invoke in the process of comparing a scientific theory to experiment, and the success of the theory in explaining experimental results cannot be taken to be literally true.

One extreme form of anti-realism derives from the subjective idealist tradition going back to Berkeley and in this context is often referred to as phenomenalism, advocated by philosophers like Stace, for example. In this view, the existence claims about the real world are wrong because what we know for sure are only the sensory perceptions they give rise to. The existence of these perceptions does not allow us to make ontological deductions about objects in the real world. This is, of course, the classical position of subjective idealism. In the context of science, this argument has been extended to say that theoretical entities that are not observable and do not even give rise to sense data are purely fictional.

Another extreme form of anti-realism is nominalism, which does not accept the existence of abstract objects, i.e., objects which are not spatio-temporal, or universals, but only accepts the existence of abstract terms. There are also more sophisticated perspectives that have come into this debate which have, in some sense, bridged the realist and anti-realist positions, like constructive empiricism. Like scientific realism, the constructive empiricism of van Fraasen also accepts a literally account of scientific theories, but only for the observable entities in the theory.[15] The focus here is shifted away from truth to empirical adequacy. A theory is viewed in semantic terms as a class of structures, and a sub-class of empirical structures in the theory finds a fit with observed phenomena. If such a fit can be obtained, the theory is empirically adequate. Van Fraasen also argues that even if an entity is unobservable but we have indirect empirical evidence for its existence, then we should believe in the existence of this entity, i.e., literally construed but not necessarily true.

More recent understanding of the knowledge question in science moves away from the linguistic description to the idea of simulation, i.e., from the semantic to the syntactic model of science, and the realist/anti-realist debate takes on new forms in this reformulation. Rather than looking for relations between classes of statements, as was done from the days of logical positivists and Popper, the simulation view of science chooses to understand how closely the simulations resemble experiments. As I said, this does not avoid the realism issue, and the question about abstract entities that appear in a simulation and their place in a knowledge system naturally appears here.

Underdetermination

Though the idea of underdetermination goes back a long time, it was first articulated in the form we recognise it now by Duhem as a problem in the empirical confirmation of theories in physics.[16] It was almost half a century later that Quine advanced it as a general problem of knowledge, not restricted to physics or even science.[17] However, the names of Duhem and Quine are jointly mentioned in discussions of the problem of underdetermination.

I will restrict my discussion of underdetermination to the philosophy of science. There are two sources of underdetermination in science. The first one comes from acknowledging that the process of empirical confirmation of theories is more complicated than it is normally understood to be. The standard view tends to ignore the guiding heuristics that are crucial for an experiment to run. Moreover, in the standard view of science, experiments should be performed without any theoretical bias, but this is not what is reflected in the actual practice of science. Other than the theoretical framework and empirical data, the set of guiding assumptions or heurisitics enter the design of the experiment. These involve an expectation of what one hopes to unravel in the experiment, and this expectation is usually based on existing theoretical prejudices. This fact is usually referred to as the theory-ladenness of science. In addition, there will be some larger assumptions made which have to do with the set of beliefs we all carry about the world we live in, for example, continuity, uniformity etc. Due to these guiding assumptions, the original theoretical hypothesis that one sought to confirm or negate in an experiment is no longer an isolated fact, but instead, one has a network or web of hypotheses, only some of which belong to the original physical set; the rest are metaphysical. So, even when confronted with a set of experimental observables, the theory remains underdetermined because a failure to correspond to the experimental fact does not discredit the original hypothesis but could be a result of one of the auxiliary hypotheses. This form of underdetermination is called holist underdetermination and, not surprisingly, it has been used extensively to argue the demise of the scientific method and the end of rationality because of the role of metaphysical assumptions. This has led to the development of several critiques of science, but I will not have the space in this chapter to go into those arguments and their counters.

The other form of underdetermination is what will interest us more in the present case, and this is what I will mean when I now use the word underdetermination. This form of underdetermination stems from the realisation that a given body of empirical data may be in agreement with two or more competing theories, and we will not be able, on the basis of the evidence, to decide between the theories. In fact, there is a strong version of underdetermination wherein theories remain underdetermined by all conceivable evidence, and a weak version where underdetermination is transient and disappears in the circumstance that new evidence becomes available in the future. The strong form of underdetermination is a serious challenge to anything like an idea of 'scientific truth' and suggests that one may have to abandon the idea of truth in favour of empirical adequacy.

Particles and strings

As mentioned earlier, Richard Dawid has put up a very vigorous case for a new approach in the philosophy of science to address string theory. Before I describe his position, I will, in this section, provide a background to the particle physics and string theory issues. Some standard references for these areas of high-energy physics are provided.[18]

Let me start with the sketch of string theory that parallels what Dawid himself presents so as to set the context for the rest of the discussion. It is best to start this discussion with the state of affairs in physics in the early 1920s. All motion, terrestrial and celestial, was understood in terms of Newton's mechanics. Two fundamental interactions of nature were known at that time: i) electromagnetism, for which Maxwell had provided a theoretical edifice and had also as a bonus with the same theory given us an understanding of light, and ii) gravitation, which was earlier understood in terms of Newton's theory but had found a new framework in Einstein's general theory of relativity. Statistical mechanics had provided a fundamental understanding of thermodynamic phenomena; Planck had discovered the remarkable phenomenon of the quantisation of energy and Bohr had developed a rudimentary model of the structure of the atom. The last mentioned achievement had raised more questions than it solved, but it also marked the journey of physics further into the microscopic domain. Not content at simply having a rough understanding of the structure of the atom, physicists wanted a complete understanding of the mechanics that were responsible for atomic structure. The answer was clear: It was not Newtonian mechanics, but an altogether new mechanics of the microscopic world called quantum mechanics. Just at the end of the remarkable decade when this understanding of quantum mechanics was put in place, two new fundamental interactions of nature were discovered: the strong and the weak nuclear reactions responsible for the dynamics that provided the binding and the breakup of atomic nuclei. In contrast to electromagnetism and gravity, these two interactions are operative only in the microscopic domain. It was important to develop theories of these interactions at the microscopic level. Further, even though electromagnetism and gravity were well-understood at large scales, one had to develop microscopic theories of these interactions, too.[19] So, quantum theories of these four fundamental interactions were needed. In fact, these theories had to be consistent with Einstein's special relativity, so relativistic quantum theories, called quantum field theories, were developed. The problem with quantum field theories were that they were plagued with infinities right from the start, but a proper treatment of these theories, in the context of what were called gauge field theories, was developed, and it was shown that with a consistent quantum-field theoretical treatment of electromagnetism, strong and weak interactions could be obtained. This theory, which is called by the rather unpretentious name, the Standard Model, has remarkably passed all the empirical tests it has been subjected to so far, and extant data from high-energy colliders to date are all consistent with the theoretical outputs of the Standard Model, which is an achievement unparalleled by any other theory in the history of science. The last bit of experimental discovery that was required to complete the Standard Model was that of the Higgs particle, which was recently discovered in experiments at CERN.

The problem, however, is in handling gravity at microscopic scales. To begin with, over a hundred years ago, Max Planck had already asked the question about the energy scale at which gravity becomes quantum—what is now called the Planck scale—and he found that it was an enormously large scale (roughly 10^{15} times the scale that the highest energy collider (CERN) operates at). Clearly, then, we are very, very far from seeing the quantum effects of gravity in experiments. But of course, we have theoretical reasons like the Bohr-Rosenfeld argument mentioned earlier, which tells us that from the point of view of theoretical consistency that gravity ought to be quantised. The obvious thing to do would be to try to quantise gravity like we did the other interactions, i.e. in a quantum field theory framework. As in the case of the other interactions, we encounter infinities, but in the case of gravity, these infinities cannot be handled using the prescriptions used in the case of the other interactions. So a quantum field theory approach to quantise gravity fails—no field theory of quantum gravity is finite.[20]

That is where the move to string theory happened, though the subject has a curious history worth recounting. String theories were first written down in the days when the Standard Model was not yet in place and a theory of strong nuclear interactions was not known. A natural picture of strings arises in the description of strong interactions, so the attempt to write down a string theory to describe strong interactions was made. However, both because of experimental and theoretical advances, the Standard Model was accepted as the correct theory and string theory was no longer needed to describe strong interactions. It was at this time that it was reinterpreted as a theory of quantum gravity because of the persistent appearance of a spin-2 particle in string theory, which in the gravity picture could be interpreted as a graviton. It turned out that the theory could also accommodate the Standard Model interactions, and so it was recast as the unified theory of all interactions.

At the same time, a new symmetry of nature, supersymmetry, was being studied in high-energy physics, and it was seen as a natural extension of the Standard Model in the sense that it is a bigger theory in which the Standard Model is subsumed and, at the same time, it cures some of the theoretical problems of the Standard Model. It turned out, quite remarkably, that supersymmetry could be naturally incorporated into the string framework, leading to what are called superstring theories. While superstring theories are unified theories of all interactions, including gravity, the mathematical consistency of the theory requires that these theories exist in a space-time with ten dimensions and not in the usual four dimensions that we live in. In order to make contact with the four-dimensional world, one has to then 'compactify' the remaining six dimensions, to make them so tiny that they are beyond any possible experimental determination. The string length and the compactification scale are both inversely proportional to the Planck scale and, consequently, unobservably tiny. At the price of having these unobservable dimensions, however, string theory brings in a host of advantages over quantum field theories. For one, the theory of quantum gravity is now finite. In fact, depending on the content and symmetries, five such consistent and finite theories could be constructed. But more importantly, the theories are specified in terms of its geometry, so there are no arbitrary parameters of the kind that appear in the Standard Model. A more detailed analysis reveals that string theories admit, in addition to strings, which are 1-dimensional objects, other objects of varying dimensionality called membranes. The theory with membranes admits of a very rich structure and reveals new relations between the five different string theories. These relations are called duality relations.

I will say a little more about these duality relations. Let us first go back to some more discussion about quantum field theories. Quantum field theories describe the interactions of matter particles, and these interactions themselves are described in terms of particles called gauge particles. To calculate any observable in quantum field theory, one would need to know the strength of these interactions, which is given by some parameters. These parameters are not constant, but depend on the energy at which the observable is measured. As it so happens, in the Standard Model, for most of the energy domain the three interactions—electromagnetic, strong and weak—are described by interaction parameters that are small. Such a domain of the theory is called the weak coupling domain. The observables of the theory admit of a mathematical series expansion in the weak coupling domain called a perturbative expansion. This is in a sense the easiest way to do a calculation in a quantum field theory. But if we move out of the weak coupling to the strong coupling domain, the interaction parameter is no longer small, and we do not have a valid perturbation expansion. Non-perturbative techniques exist but are fewer and difficult, but these are the techniques that have to be used to extract useful information from the theory in the strong coupling domain. One known non-perturbative technique is duality wherein, given a theory *T*, one writes down a theory *T'* such that the interaction parameters in the two theories are inversely related. So the strong coupling domain of *T* becomes the weak coupling domain of *T'*, and it is possible then to use perturbative calculational strategies in *T'* to extract information about the strong coupling domain in *T*. The difficulty is to know the dual theory *T'* given the original theory *T*. The other thing is that *T'* may not be the same as *T*; in fact, in most of the cases, it is a completely new theory with new theoretical entities. As mentioned, duality has been studied in the context of field theories, but also in the simpler context of spin systems in statistical mechanics (which, in any case, are very much like quantum field theories). It turns out that in string theories, duality becomes a very powerful tool. There are several duality relations that relate the strong and weak coupling domains of different string theories and, in the end, the five distinct string theories one started with all turn out to be the manifestation of a fundamental theory called *M* theory. Even more remarkably, another example of a duality relation in string theory relates a string theory in a particular higher-dimensional space-time to a four-dimensional quantum field theory (of course, a very specific kind of a quantum field theory). Thus, duality in string theory has emerged as a very powerful tool to find equivalences between apparently very dissimilar theories and even between a specific string theory and a quantum field theory!

String theory and issues in the philosophy of science

Having provided the background necessary to understand the issues involved, I will now summarise Richard Dawid's position. In this section, I will present the main strands of his argument as presented in his book and his papers. Here, I will only state his position and reserve my response for the next section.

Dawid argues that string theory is deserving of a non-empirical assessment because of the following reasons:

i) The no-alternatives argument,
ii) The argument of explanatory coherence,
iii) The argument from the success of other theories in the research programme called the meta-inductive argument.

The no-alternatives argument that Dawid puts forward is that string theory is a quantum theory of gravity that also unifies gravity with the other interactions of the Standard Model. Other approaches to quantum gravity, like canonical quantum gravity, do not provide this theoretical framework of unification of gravity with other interactions, and the Standard Model, as we already mentioned, does not address gravity.

He also mentions an argument due to Polchinski, but it is worth describing Polchinski's argument in some detail rather than merely mentioning it, as Dawid does. Polchinski argues that in attempting to come up with a completely new theory of space-time at the Planck scale, one may depart from the usual quantum mechanics and, in addition to the usual position-momentum uncertainty relations of quantum mechanics, one may posit position–position uncertainty relations.[21] Such a possibility may be realised by treating position coordinates as matrices instead of simple numbers. He considers a system of *N* non-relativistic particles and writes down a Hamiltonian to describe it. The Hamiltonian at large *N* contains the commutators of the different components of the matrices corresponding to the position coordinates summed over all the particles. Of course, at energies much smaller than the Planck scale, the matrices need to commute so as to recover the usual quantum mechanics. To ensure this, one needs to supersymmetrise this Hamiltonian, and so we end up with a supersymmetric quantum mechanics system. In a classic paper by Banks, Fischler, Shenker and Susskind, it was shown that such a supersymmetric quantum mechanics Hamiltonian describes *M* theory in a certain limit called the infinite momentum limit, i.e., this simple Hamiltonian describes the theory underlying the five different string theories viz., *M* theory.[22] Thus, starting from the simple motivation of modifying quantum mechanics at the Planck scale, one ends up with string theory, and Polchinksi claims that this is a proof that the only physics possible at the Planck scale is string theory. Dawid then takes Polchinski's somewhat hyperbolic claim to suggest that in fact, string theory has no alternatives.

The second argument Dawid advances is that of the unexpected explanatory coherence of string theory. He argues that gravity emerges from string theory in a natural way and gauge theory, unification and supersymmetry are also incorporated in string theory. That it brings these different theoretical ideas together is an argument in favour of it.

The empirical success of the Standard Model, what Dawid views as a precursor to string theory, is taken to be another argument in favour of string theory. The Standard Model was proposed on theoretical grounds but has been empirically tested to a very impressive level of accuracy. The prediction of the new particles that the model required and their discovery in experiments like the W and Z bosons and the Higgs boson are successes not just of the Standard Model, but a success of the theoretical programme of high-energy physics. To quote his meta-inductive hypothesis: 'Scientific theories which are developed in the research programme of high-energy physics in order to solve a substantial conceptual problem, which seem to be without conceptual alternative and which show a significant level of unexpected internal coherence tend to be experimentally successful once they can be tested by experiment'.

It is using these three points made above that Dawid makes a claim for non-empirical theory assessment. He admits that non-empirical assessment is not on the same epistemic footing as empirical assessment, but argues that it is worthy of consideration. It is not a criterion for theory assessment that should be completely ignored, and the superior status that we accord to empirical assessment is exaggerated and misplaced. Dawid imagines a hypothetical world of superior beings who have the ability work out all the alternative theories and every such theory is worked out in a mathematically and logically coherent manner. In this world, if the scientists were to find that every single scientific theory to which they could not find any alternative has eventually turned out to be verified empirically then for such theories non-empirical assessment would be a very significant tool. For Dawid, in such a strange world, the belief in non-empirical assessment of theories would have the same status as we have in empirical theories. We use well-tested theories of science to build aeroplanes and bridges, he says, and it is because of our belief in empirical testing that we are willing to use these bridges and planes, even knowing that the failure of this belief may come at the cost of our lives. Dawid's claim is that in the hypothetical world that he considers, people would believe in non-empirical testing (based on the no-alternatives assessment) in quite the same way: They would literally let their lives depend on it. He says that just as a person in our world who refuses to use a bridge or a plane because of the lack of belief in empirical testing would be considered weird, in this hypothetical world, a lack of belief in non-empirical theory assessment would be considered weird. Invoking this world, Dawid wants to drive home his point that the difference (in our own world) between the empirical and non-empirical modes of theory assessment is not one of kind, but only one of degree.

Having made a case for taking non-empirical theory assessment seriously, Dawid demands that it not only be applied to the case of string theory, but that it has actually been the way high-energy physics has developed over the last fifty years. In particular, he claims that the belief in the Standard Model before the empirical discovery of its various elements is an example of non-empirical assessment. The fact that these elements of the Standard Model were then empirically determined makes the case for a similar assessment of string theory stronger. Dawid mentions the discovery of the Higgs boson and the top quark as instances from the Standard Model that support his argument. In short, his argument is that there were theoretical reasons to believe that the top quark and the Higgs boson existed in the Standard Model and their existence could be predicted from arguing from symmetry principles within the Standard Model and some inviolable features of quantum field theory. So the belief in the existence of these particles was held by the community long before their discovery.[23] Dawid argues that this increases our confidence in using a non-empirical assessment procedure to evaluate string theory.

Finally, we will discuss one more point that Dawid raises in proposing. This has to do with the role played by dualities in string theory. We will present Dawid's point briefly. It is suggested by him that dualities play a crucial role in string theory, and they are substantively different from those we encounter in field theories. For example, he argues that in field theory, duality relations simply imply a relation between a weak and strong coupling theory, and one posits it only as a convenience, i.e. to handle the problem that the strong coupling theory is intractable for, and one works with its dual weak coupling theory. He goes on to say that in string theory, duality enters differently and is hitting right at the heart of the realism/anti-realism debate, or that duality is not tenable at all within a perspective of ontological realism.

A new philosophy of science?

We started this chapter with a general reflection on science, its method and its practice. Or we should say, more correctly, the sciences, their methods and their practices. Our singular phrasing was in anticipation of the fact that we are not attempting to talk about a method of all sciences but only of physics, and that too, largely of high-energy physics. Having said that much of what comes out of these debates, even if they are largely about high-energy physics, has implications for all areas of science, including the social sciences. At the end of reviewing the different positions that have been adopted in this recent debate on the epistemological and ontological status of high-energy physics and string theory, we are left with the overwhelming question: Have we then genuinely reached a point in high-energy physics that we now need a new philosophy of science? In particular, is that new philosophy something along the lines that philosophers of science like Richard Dawid advocate?

We mentioned earlier that the Standard Model of particle physics, the theory of the three fundamental interactions in nature (electromagnetism, strong interactions and weak interactions), is a complete theory, in agreement with all extant data from high-energy physics experiments and at a level of accuracy that is unprecedented.[24] There are a couple of empirical indications suggesting theories beyond the Standard Model, which are the observation of neutrino oscillations and the existence of dark matter in the universe. The first, while it is accommodated with some modification of the Standard Model, does definitely point to some unknown new physics origin for the mass of the neutrino. None of the Standard Model particles are good enough candidates to explain the dark matter puzzle. So, if this puzzle genuinely has a particle physics origin (and there is a possibility it may not), then it could come from one of several theories that go beyond the Standard Model. But there are some theoretical reasons that again suggest that the Standard Model is not a complete story, like gauge hierarchy[25] or the hierarchy of masses of the observed matter particles. So while at the present juncture, the Standard Model is in agreement with all empirical evidence, the high-energy physics community believes that there will be new physics that experiments will discover and that, too, in the very near future. The Large Hadron Collider at CERN, which discovered the Higgs boson a couple of years ago and established the Standard Model as the correct theory of the microcosm, will be collecting and analysing data for the next few years. In these years, it is hoped that it will find some evidence for new physics, either of the kind already anticipated by theorists or something completely new and unexpected. If this does not happen and no new physics that falsifies the Standard Model is discovered at the Large Hadron Collider, then it is likely to be very difficult for the community to convince funding agencies to release funding for new colliders, and that will seriously impact the progress of high-energy physics and, indeed, of our understanding of the micro-world. It is, therefore, a mixed feeling of achievement, anticipation and anxiety that besets the high-energy physics community today. In this field of high-energy physicists, a subset who work on string theory have been working on not just the theory of the three interactions, but on a theory which also incorporates gravity. The theory is a natural setting for quantum gravity, and its natural energy scale is enormously higher than what current experiments are able to reach. Consequently, the theory is disconnected from the world of high-energy experiments. An already-anxious larger community that is desperately seeking new observable phenomena in current experiments is not very excited about final theory claims of a subject that will not make contact with experiment.

This problematic situation has given rise to two reactions within the community. On the one hand, it has given rise to string model builders or string phenomenologists, i.e. those who strive to make models derived from or based on string theory to make contact with real-world experiments, while, on the other hand, it has created another sub-group of researchers who are even less connected with anything to do with the experimental world and pursue their subject for purely theoretical and mathematical reasons. Part of this has to do with the fact that models derived from or based on string theories usually contain assumptions and guesses and the introduction of more parameters not contained in the original theory. With much of the elegance and beauty that drive the mathematically minded practitioners of the subject forsaken, they are driven even further away from the real world. There is simultaneously a worry that this practice of pure theory cannot go on forever, and that eventually, they will face problems in getting to fund this kind of research. Thus, the high-energy physics stage is divided into two, with actors on both sides ridden with serious existential issues but also involved in a struggle to occupy centre stage. It is within this context of present-day high-energy physics that the debates discussed in this chapter need to be seen.

Let us now go back to Dawid's arguments for advocating a non-empirical assessment of string theory. The first argument was that a theory that has no alternatives is a candidate for non-empirical assessment, and string theory is one such candidate because there are no alternatives to string theory. As a possible alternative, Dawid mentions loop quantum gravity, but dismisses it because loop quantum gravity is only a theory of gravity, whereas string theory is a theory of all interactions. So, according to Dawid, loop quantum gravity is not an alternative. This argument will work only if unification is a logical necessity, but this is obviously not so. Let us state the case more clearly. String theory is a theory of gravity and the other three interactions that are described by the Standard Model. As a theory of quantum gravity, other theories like loop quantum gravity should be thought of as alternatives, whereas for the three other interactions, string theory is an alternative to the well-tested Standard Model. The problem is that string theory is a theory of these latter three interactions only in ten dimensions, and one does not a have a four-dimensional version of the theory with the right features and, particularly, the quantitative predictive power that the Standard Model has. So, while string theory is an alternative to the Standard Model, it is, at present, a very poor alternative. As a theory of quantum gravity, a theory like loop quantum gravity is also an alternative to string theory. That the quantum gravity sector is not empirically accessible is a problem that plagues string theory and loop quantum gravity alike. At a theoretical level, proponents of each theory level criticisms at the other, so eventually, only an empirical test can put an end to these theoretical disputes. In this context, it is important to remember the history of the original Kaluza-Klein model. Dawid's arguments would suggest that it is a superior theory to Einstein's theory of gravity and Maxwell's theory of electromagnetism. The reality is that the unification of gravity and electromagnetism as proposed by Kaluza and Klein was unviable. So the physical idea of unification cannot be elevated to a philosophical principle. In the case of Maxwell's theory, unification works, but it does not in the Kaluza-Klein case. So unification is an assumption of physical theories that needs to be tested in experiments.

I also want to quickly comment on the argument of Polchinski about there being no alternative to string theory. The core of this argument, as we said before, comes from the well-known paper by Banks et al. Polchinski turns this argument on its head and seemingly ends up with a string theory by starting with rather general assumptions at the Planck scale which did not assume anything about string theory. The starting point of assuming position–position uncertainty relations is not the most general thing one can do. It was chosen because one had the benefit of the result of Banks et al. To claim from such an argument that there are no alternatives to string theory is vacuous.

While discussing alternatives, it is also important to talk about supergravity theories. When supersymmetry is made a space-time-dependent symmetry, then it is called a supergravity theory because gravity arises naturally in such a theory. Supersymmetry is one way of taming the infinities of quantum field theory, so in supergravity theory, it was hoped that one could obtain a finite theory of gravity. For this one needs a supergravity theory with maximal supersymmetry: what is called N = 8 supergravity.

This theory was investigated vigorously in the late 70s and early 80s with the hope that it may yield a finite field theory of quantum gravity. In the early 80s, however, there were indications based on extrapolations of calculations in N = 8 supergravity that this theory is not finite. What was initially proposed as a conjecture very quickly took the form of a certainty, and this was what led to the firm belief in the community that N = 8 supergravity is not finite and, as a consequence, no field theory of quantum gravity is possible. However, these calculations are done using perturbative techniques, which as we mentioned before, involves doing series expansions with a small parameter. In the 80s, these techniques could be used to calculate only up to some fairly low orders in the series expansion. More recently, powerful techniques for amplitude calculations in field theories have been used to calculate higher-order coefficients in N = 8 supergravity and surprisingly, these calculations suggest that N = 8 supergravity may be a finite theory.[26] In the last few years, these calculations have been extended and crosschecks have been made and the claims of finiteness are on a firmer footing now. So now a field theory of quantum gravity becomes viable again and is a serious alternative to string theory.

Dawid also argues about the consistency and coherence of string theory. While this is true, not everything about this is miraculous. For example, one can argue that the Standard Model is affected by theoretical problems like the gauge-hierarchy problem and requires supersymmetry to cure it. Supersymmetry, on the other hand, is incorporated in superstring theories. So it would seem that string theory neatly incorporates something that would help address one of the problems of the Standard Model. However, to address the hierarchy problem of the Standard Model, supersymmetry should be manifest at the energies at which we are now doing experiments. If we do not find supersymmetry at the energies that the Large Hadron Collider is now operating at, then we will lose faith in supersymmetry as a solution to the gauge-hierarchy problem. But the supersymmetry that string theory requires can be at a very high-energy scale. The non-observation of supersymmetry at low scales will have no impact on string theory. Nor is string theory vindicated if supersymmetry is discovered at the Large Hadron Collider.

The argument based on meta-induction is also specious. Let us say that in the next few years, we have definite evidence for low-energy supersymmetry. That evidence would also rule out one serious alternative to low-energy supersymmetry, which is called Technicolor. So one theoretical approach that had been held as a belief in the community gets confirmed by empirical data while another theory gets falsified, showing us that the meta-induction argument is false. The meta-induction argument relies heavily on the no-alternatives argument, which, as we have just demonstrated, does not hold. Dawid also presents a description of the Standard Model as though it was a theory with no alternatives. This is not true at all: There have been alternatives to the Standard Model, the Georgi-Glashow model,[27] for example. It was the experimental discovery of neutral currents that selected out the

Standard Model between these alternatives. Similarly, before the Higgs boson and the top quark were discovered, theorists did study variants of the Standard Model without these particles to see whether they are viable. The belief in the existence of the Higgs or the top before their discovery came from the fact that these other models were not viable in reproducing other experimental results that gave indirect information about these particles. The increase in belief again was due to empirical and not meta-theoretical reasons.

It is also important to mention that string theory has been studied by the philosophers Johansson and Matsubara as a research programme from a Lakatosian perspective, and they have found it to be a degenerative research programme.[28] Their results are not inconsistent with a similar study done earlier by Cartwright and Frigg.[29]

Finally, I want to discuss the issue of duality and the demise of ontological realism. I will do so briefly for want of space and also because I have written about this elsewhere.[30] While I agree with Dawid that dualities make an ontological realist perspective unviable, I do not agree with the conclusions that he draws using this argument. It has also been argued that there is a tripartite unification that saves realism in dual theories, i.e. an underlying theory that is the deeper reality of the dual theories.[31] It is not clear that an underlying theory always exists, so this seems just a desperate attempt to save realism from the attack of dualities. Somehow Dawid uses this to argue for the special status of string theories, i.e., they deserve a special epistemic status because even at the ontological level, string theories are special. This is an erroneous conclusion to my mind. Dualities exist even in field theories, and while they are not relations between theories as disparate as the ones in string dualities, they still bring into question the issue of ontological realism. In fact, they become better testing grounds for the conceptions of realism and anti-realism because they are simpler than string theory dualities. Even simpler are the Kramers-Wannier dualities in statistical spin systems. An Ising model in two dimensions has a dual description also in terms of a two-dimensional Ising model, i.e. it is self-dual. Nevertheless, the dual model has features the original model does not have. In particular, one can start with a purely ferromagnetic theory, but the dual will have a mixture of ferromagnetic and anti-ferromagnetic couplings. This gives rise to the phenomenon of frustration and has been extensively used to model spin glasses.[32] These features are not discernible in the original theory, and so the issue of ontological realism is the same as in string theories. If at all, in the case of statistical systems and field theories, the case against realism is even stronger because one is dealing with observables directly related to empirical phenomena.

Reflections

I have deliberately called this last section Reflections rather than the usual Conclusions because I am going to use these few lines to comment on conclusions in science. Science, in general, and physics, in particular, is largely a discourse of conclusions. It suffers from an anxiety of wanting to get to a point where it can be said, 'So we conclude . . .'. The dream of a final theory is this anxiety perversely transformed into an utopian dream, and again it is due to this anxiety that the Standard Model itself is seen as a finished product, even though high-energy physicists know that some calculations that are too difficult and have not been done and checked against empirical results are still believed to be 'verified'. Maybe the clue to new physics lies there, in the undone calculations and unperformed empirical tests even at the level of the Standard Model. Instead, in the search for ultimate answers, these smaller issues fall by the wayside. It may be a long time before we have an empirically viable theory of gravity. If in the meantime, we also attend to other research issues that we have ignored, then it will help alleviate some of the 'final theory' anxiety. We are nowhere closer to wanting a new philosophy of science that rejects empiricism and falsifiablity, at least not now.

Notes

1 Boltzmann, L., 'On statistical mechanics' in *L. Boltzmann: Theoretical Physics and Philosophical Problems*, B. McGuinness (ed.) (Dordrecht: Reidel, 1974), p. 163.
2 For a comprehensive discussion of this history, see Wilholt, T., 'When realism made a difference: The constitution of matter and its conceptual enigmas in late 19th century physics', *Studies in History and Philosophy of Science Part B: Studies in History and Philosophy of Modern Physics* 39(1) (2008), pp. 1–16.
3 Dawid, R., 'Theory assessment and final theory claim in string theory', *Foundations of Physics* 43(1) (2013), pp. 81–100.
Dawid, R., 'High energy physics and the marginalization of the phenomena', *Issues in the Philosophy of Physics* 33(1) (2010), pp. 165–206.
Dawid, R., 'Scientific realism in the age of string theory', *Physics and Philosophy* 11 (2007), pp. 1–32.
4 Dawid, R., *String Theory and the Scientific Method* (Cambridge: Cambridge University Press, 2013).
5 Carroll, S., 'What scientific idea is ready for retirement', in the *Edge* (2014): https://edge.org/response-detail/25322
6 Ellis, G. and Silk, J., 'Defending the integrity of physics', *Nature* 516 (2014), pp. 321–23.
7 Hossenfelder, S., 'Post-empirical science is an Oxymoron', in *Back Reaction* (2014): http://backreaction.blogspot.ch/2014/07/post-empirical-science-is-oxymoron.html
8 It is of interest to note in this context that in the concluding talk of the Strings 2014 Conference held in Princeton, David Gross actually presented a slide showing Dawid's book cover and talked about the new method of science based on non-empirical testing and concluded, in the manner of a surgeon reporting on his patient, that 'String Theory is alive and healthy!'
9 Pigliucci, M., 'Sean Carroll, edge and falsifiability', *Rationally Speaking* (2014): http://rationallyspeaking.blogspot.ch/2014/01/sean-carroll-edge-and-falsifiability.html
10 Popper, K., *Logic of Scientific Discovery* (London: Routledge, 1992) (First published in German (Vienna: Springer, 1935) and in English (London: Hutchinson, 1959)).

Popper, K., *Realism and the Aim of Science* (New Jersey: Rowman and Littlefield, 1983) (First published in 1956).

11 Hume, D., *Enquiries Concerning Human Understanding and Concerning the Principles of Morals*, Edited and revised by P. H. Nidditch from the 1777 edition edited by L. A. Selby Bigge (Oxford: Clarendon Press, 1975).

See also, Carnap, R., *The Philosophical Foundations of Physics* (New York: Basic Books, 1966).

12 Popper toys with the idea of using the term 'perception statement' but to avoid any other connotation to this term, he prefers to use the simpler 'basic statement'.

13 Ref. 9, p. 42 and p. 50, respectively.

14 For reviews, see, Brock, S. and Mares, E., *Realism and Antirealism* (Chesham: Acumen, 2007).

Rothbard, D., *Science, Reason and Reality: Issues in the Philosophy of Science* (Texas: Harcourt Brace and Company, 1998).

Psilos, S., *Scientific Realism: How Science Tracks Truth* (London: Routledge, 1999).

15 van Fraassen, B., *The Scientific Image* (Oxford: Oxford University Press, 1980).

16 Duhem, P., *The Aim and Structure of Physical Theory*. Translated from 2nd edn. by P. W. Wiener (Princeton, NJ: Princeton University Press, 1954), originally published as *La Théorie Physique: Son Objet et sa Structure* (Paris: Marcel Riviera & Cie., 1914).

17 Quine, W. V. O., 'Two dogmas of empiricism', Reprinted in *From a Logical Point of View*, 2nd edn. (Cambridge: Harvard University Press, 1951), pp. 20–46.

18 For technical accounts of the Standard Model, see:

Cheng, T. P. and Li, L. F., *Gauge Theory of Elementary Particle Physics* (Oxford: Oxford University Press, 1984).

Aitchison, I. J. R. and Hey, A. J. G., *Gauge Theories in Particle Physics*, Vols. I and II (New York: Taylor and Francis, 2003).

For a non-technical introduction to the Standard Model, see:

Oerter, R., *The Theory of Almost Everything: The Standard Model, the Unsung Triumph of Modern Physics* (New York: Pearson, 2006).

For technical accounts of string theory, see:

Green, M. B., Schwarz, J. H. and Witten, E., *Superstring Theory*, Vols. I and II (Cambridge: Cambridge University Press, 1987);

Polchinski, J., *String Theory*, Vols. I and II (Cambridge: Cambridge University Press, 1998).

For a non-technical introduction to string theory, see:

Greene, B., *The Elegant Universe: Superstrings, Hidden Dimensions, and the Quest for the Ultimate Theory* (New York: W. W. Norton, 1999).

For a technical account of supersymmetry and supergravity, see:

Binetruy, P., *Supersymmetry: Theory, Experiment and Cosmology* (Oxford: Oxford University Press, 2006).

For a non-technical introduction to supersymmetry, see:

Kane, G., *Supersymmetry: Unveiling The Ultimate Laws of Nature* (New York: Basic Books, 2000).

19 Other than empirical exigencies, there was also an issue of theoretical consistency in demanding that these classical interactions be quantised and an argument due to Bohr and Rosenfeld.

20 Or so we believed.

21 Polchinski, J., 'Quantum gravity at Planck length', *International Journal of Modern Physics, A* 14 (1999), pp. 2633–58.

22 Banks, T., Fischler, W., Shenker, S. H. and Susskind, L., 'M theory as a matrix model: A conjecture', *Physical Review D* 55 (1997), pp. 5112–28.

23 The top quark was discovered during the Tevatron experiment at the Fermilab in Chicago in 1995 and the Higgs boson was discovered during the Large Hadron Collider experiment at CERN in Geneva in 2013.

24 This does not mean that every observable in the Standard Model has been calculated and verified. See the last section of this chapter for a discussion of this issue.

25 This is a theoretical problem related to the existence of the Higgs boson in the Standard Model. The Higgs boson has no spin (all the other particles have spin-1/2 or spin-1) and spinless particles, called scalars, have pathological quantum properties. One way to cure their illness is to invoke a new theory: supersymmetry.

26 Bern, Z., 'Is N=8 supergravity ultraviolet finite?', *Subnuclear Series* 45 (2009), pp. 25–40.
Kallosh, R., 'An update on perturbative n=8 supergravity', *Los Alamos Archives: arXiv hep-th* 1412.7117 (2014).

27 Georgi, H. and Glashow, S. L., 'Unified weak and electromagnetic interactions without neutral currents', *Physical Review Letters* 28 (1972), pp. 1494–96.

28 Johansson, L. G. and Matsubara, K., 'String theory and general methodology: A mutual evaluation', *Studies in History and Philosophy of Modern Physics* 42 (2011), pp. 199–210.

29 Cartwright, N. and Frigg, R., 'String theory under scrutiny', *Physics World*, September 2007, pp. 14–5.

30 Sridhar, K., 'The method of science and its discontents', in *Breaking the Silo: Integrated Science Education in India*, A. Dhar, T. Niranjana and K. Sridhar (eds.) (2015).

31 Read, J., 'The interpretation of string-theoretic dualities', *Philosophy of Science Archives*: http://philsci-archive.pitt.edu/11205/ (2014)

32 Savit, R., 'Duality in field theory and statistical systems', *Reviews of Modern Physics* 52 (1980), pp. 453–547.

9 Cognitive constraints on the perception of reality

Avi Chaudhuri

The eye sees only what the mind is prepared to comprehend.

Robertson Davies

Our perception of the physical environment requires two very different processes. The first step is the acquisition of sensory signals through the various receptors that have evolved over eons, such as the eyes, ears, nose and so on. Psychologists refer to the sequential analysis of sensory signals as *bottom-up* processing. A key element in this construction is that sensory systems are engaged in information processing of only environmental signals and do so by piecing together small chunks of the stimulus to eventually create a perception of the stimulus. Sensory processing, according to this theory, begins with the stimulus itself and is then processed in a direct manner through successive stages.[1] One school of thinking contends that all the necessary information required by our brain can be recovered from the environment to produce a mental representation. Perhaps Einstein would have belonged to that school.

British psychologist Richard Gregory proposed a completely different theory of perceptual processing. According to Gregory, sensory perception is not just the mere outcome of sequential bottom-up processing, but rather depends strongly on the past experiences, contextual information, inference, knowledge, mental state and even the cultural background of the individual.[2] Thus, perception is the outcome of *top-down* processing in which cognitive factors play key constructive roles in generating perceptual awareness. Unlike the direct and automated manner of bottom-up processing, the top-down theory proposes that perceptions are constructed in an indirect manner. The brain can only make sense of the incoming sensory signal by relying on past experience and prior knowledge. The *constructivist* theory of Gregory therefore represents a fundamentally different way of thinking about perception, and it is this theory to which Tagore would presumably have belonged.

In this chapter, I will explore the various top-down factors that can affect sensory perception and explore the cognitive mechanisms that play key roles in that process. The very notion of *reality* is discussed within the context of what it means to perceive something and the way our mind creates that internal reality. My goal will not be to cover the full breadth and details of all cognitive factors that influence perception, which can easily fill a book on its own. Rather, my discussion will be aimed at highlighting the key factors that are involved in top-down processing. I begin with a discussion on the role of attention and how it can modulate sensory function. The roles of various other high-level cognitive factors, such as motivation, emotion and culture are then taken up in the context of guiding the *perceptual set*.

Attention

The process by which we select or concentrate on one aspect of our sensory environment is known as *attention*. A classic explanation of attention was given by William James in 1890:

> Every one knows what attention is. It is the taking possession of the mind, in clear and vivid form, of one out of what seem several simultaneously possible objects or trains of thought. Focalization, concentration, of consciousness are of its essence. It implies withdrawal from some things in order to deal effectively with others and is a condition which has a real opposite in the confused, dazed, scatterbrained state which in French is called distraction, and Zerstreutheit in German.

We constantly participate in and experience the phenomenon of attention. At times, we may engage a global attentional process through which a diffuse and general gathering of sensory information is undertaken, whereas more often, we are concentrating on a particular location in space, a specific object or even a particular property of that object such as its shape, colour or texture. We also know from everyday experience that we are capable of extracting much greater information about objects and their properties when we selectively allocate attention. For this reason, *attention* is generally regarded as a filtering process, allowing the human brain to discard irrelevant sensory signals from the real world and to focus instead on those deemed to be of interest or importance at any given moment.[3,4,5] Our perception of reality can be affected in significant ways through the influence of attention, to the point, for example, where we may not even see a gorilla walking across a basketball court.

Selective attention

The emergence of experimental psychology in the 19th century began an era of intense research on attention. A key problem at the outset concerned the precise way in which the influence of attention on perceptual processes can be quantified. On the one hand, the early German scholars were struggling to understand how sensory phenomena can be measured, and yet now comes into the mix the further element of attention whose very properties were quite unknown at the time. The solution proposed by Wilhelm Wundt, widely considered to be the founder of modern psychology, was to assess mental processing speed indirectly by measuring the time needed to react to a signal. Wundt discovered that reaction times were shorter and more consistent if observers were given a heads up about a stimulus that was about to appear than if the warning was absent. From this finding sprang the idea that mental processing speed and accuracy can be enhanced (or reduced) depending on the goals and consequences that are in play at that very moment. A scientific approach to examining the role and impact of *selective attention* was now born, and to this day remains as one of the most widely studied phenomena in cognitive science.

Bottom-up versus top-down attention

We know from everyday experience that we orient our eyes towards objects of interest so that their image is cast centrally on our eyes, allowing our visual brain to allocate much greater computational resources to scrutinising the image in detail. However, what is it that draws our eyes to the stimulus in the first place, or put another way, what captures our attention toward that object? One way that our attention can be drawn is by the object itself—whether by its sudden movement, or its unusual feature or perhaps from the sound it makes. There are many things associated with real objects in the physical world that are constantly drawing our attention to them so that a greater analysis can be undertaken. These are all examples of stimulus-driven attention and lead to an automatic response in behaviour through bottom-up processing.[6]

A completely different situation arises when the impulse comes from within, that is, when we ourselves choose to orient our attention toward an object in a voluntary manner. The mere act of reading the words on this page or scanning the environment, say, at a concert or sports event represent examples of goal-driven attentional control where we actively choose to select the stimulus from a rich sensory environment.[7,8] These are executive decisions that we continuously make due to shifts in interest or through demands placed on our sensory systems from ongoing behaviour—such as reading, driving, hunting etc.—and therefore represent top-down attentional control.

Spatial attention

The examples discussed above, whether due to bottom-up or top-down control, lead us to conclude that attention is allocated to distinct regions of space to coincide with the object of interest. However, does this necessarily mean that we always have to move our eyes in order to allocate attention to that point in space? Let us consider the following neat experiment by American psychologist Michael Posner illustrated in Figure 9.1. Observers are seated in front of a computer monitor and are asked to stare at the central fixation point (marked by an x) and to ensure that they do not deviate their eyes from this position. In the next frame, a cue is presented by way of an arrow facing either horizontal direction. In the final frame, a stimulus appears either to the right or left of the fixation point. The subject must hit a button as soon as this stimulus is detected.

Posner made a fascinating discovery. The reaction time of his subjects was significantly lower when the stimulus appeared in the same location as pointed by the arrow—the so-called *valid cue* condition—than when it appeared in the opposite direction (*invalid cue* condition). Keeping with Wundt's view, it would appear that mental processing is more efficient when the observer is cued to where the stimulus is to appear than when it appears in a different location. Why should this be the case? Posner reasoned that the cue produces goal-directed attention to the point in space indicated by the arrow, which in turn enhances mental performance for stimulus detection. Performance is slower when the stimulus appears in a non-attended location. What is perhaps most surprising is that the allocation of attention in this experiment occurs without any overt movement of the eyes. Recall that the subject was instructed to strictly maintain fixation on the central spot, and therefore the cue only produced an internal spatial shift in attention.

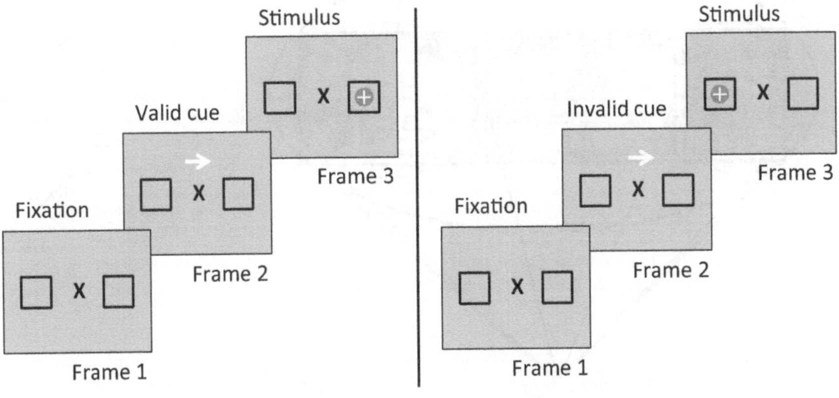

Figure 9.1 Allocation of spatial attention can be studied by cuing human observers to the location where a stimulus is about to appear. When a valid cue is presented (left panel), the reaction time for stimulus detection is shorter than when an invalid cue is given (right panel).

The attentional spotlight

The conclusion reached by Posner in his classic experiment was that allocation of attention is a cognitive process that need not necessarily be tied to eye movements. Simply put, we can be looking at one thing and attending to something else. We engage in such behaviour all the time—for example, looking briefly at the GPS monitor during driving while attending to the front of the car, or attending to the movements of nearby teammates while playing a sport or perhaps an all-too-common scenario, looking at a professor during a lecture, all the while attending to an attractive classmate in the periphery.

Although our attentional path usually coincides with gaze direction, it is also true that we spend a lot of time veering our 'mind's eye' away from what we are staring at. This feature of the attentional system has been likened to that of a spotlight, as illustrated in Figure 9.2. An object that falls within the *attentional spotlight* is then selected for greater processing. It is believed that our attentional spotlight is continuously scanning the environment, an evolutionary outcome that allowed us to cognitively survey the landscape in order to select and more closely engage with objects that had survival value or posed a threat. This leads to the question: can the size of the *spotlight* be changed? There is some evidence that the attentional beam is not fixed but can vary in size by zooming in and out upon a target.[9,10,11] The aptly named *zoom-lens* model of attention implies that the width of the spotlight can be altered and that attention can be focused depending on the stimulus requirements, such as a single letter or an entire word in a sentence (right panel in Figure 9.2).

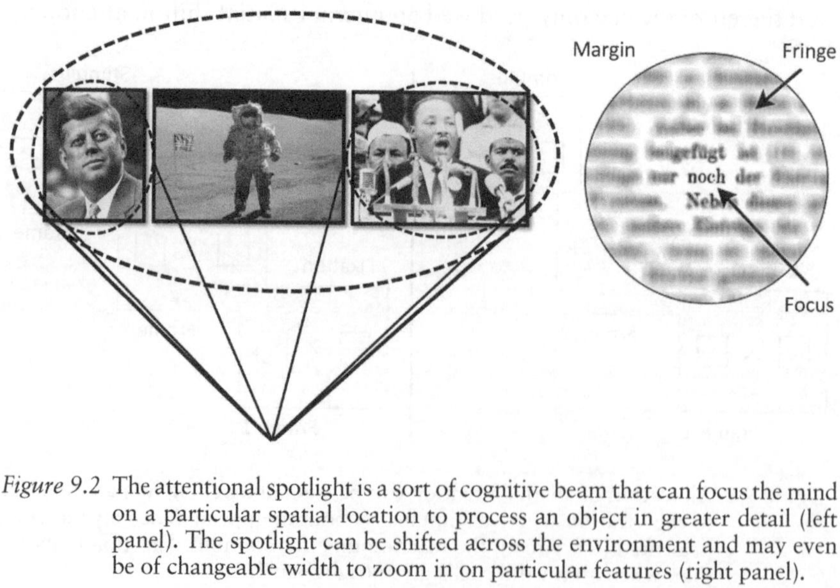

Figure 9.2 The attentional spotlight is a sort of cognitive beam that can focus the mind on a particular spatial location to process an object in greater detail (left panel). The spotlight can be shifted across the environment and may even be of changeable width to zoom in on particular features (right panel).

Divided attention

Our discussion above raises an interesting question: Is there just one spotlight of attention, or can there by multiple spotlights? Consider the well-known fact that we are capable of simultaneously attending to different objects, such as watching TV and listening to a friend talking next to us or gazing out the window at a passing car while following the lyrics of a favourite song playing on the radio. It is possible that in such cases, what we are actually doing is quickly shifting a single attentional spotlight to multiple locations or events back and forth. There is evidence, however, that attention can actually be split between two or more locations and that our attentional system may actually operate as multiple spotlights at any given time.[12,13]

If the multiple attentional spotlight theory is correct, then how good are we in tasks that require divided attention? That depends on a number of factors. One of them is task familiarity. Performance increases with practice to the point where certain tasks become nearly automatic. This means that although attention is still required in such cases, less of it needs to be allocated so that other stimuli can be attended to simultaneously. A good example again is the driving scenario. Most of us have experienced the great amount of concentration needed when first learning to drive, and yet with time and greater familiarity, many of the routine tasks of driving become automated and therefore demand less of our attentional resources. Whether in the modern workplace or our daily personal lives, *multitasking* has become a common requirement thanks largely in part to our new digital world.

A number of other factors also influence our ability to split our attention. These include the difficulty of the task, mental state (such as alertness and emotion) and even the intelligence of the individual.[14,15] Task difficulty plays a particular important role. If a particular task is so demanding of our attention, then divided attention becomes nearly impossible because all of our finite attentional resources are then allocated to that one task. A familiar example is the common and dangerous practice of talking on a cell phone, or even worse, texting while driving. A great amount of attention is required for this behaviour, especially texting, to the point where attention is completely taken away from the events unfolding on the road ahead.

Object-based attention

There is an entirely different aspect of attention that is dedicated around the object itself. To understand this, let us consider the following experiment. Figure 9.3 shows a set of words that have been printed in different colours. In the first part of the experiment, read aloud through the entire set of words as printed. This should be pretty simple. Now, read out the words based on the colour of the text rather than what is actually written. The first two lines are rather straightforward, but then what happens in the remaining lines? Most people will fly through the first two lines but then struggle with the remaining ones. This is because a conflict arises between what is actually written—something that we process automatically from our lifelong experience of reading text—and the content of the text in terms of its colour appearance. When we are forced to process only the colour in this task, performance deteriorates rapidly because we are so unaccustomed to reading in this manner.

But there is something else going on here. If a set of geometric patterns or other tokens were used instead of words, then this task would be straightforward. Our difficulty in calling out the words in 3 occurs due to a conflict arising from the automaticity of word reading and the task at hand—i.e., devoting attention to a particular feature, in this case, the colour of the word. The meaning of the word and its colour collide, and we therefore have to reconcile that in our mind as we plow through the text. The curious part of this experiment is that our attention is not being conflicted in a spatial manner. In other words, we do not have to divert our attention between two separate locations or two different events (such as driving and texting at the same time). Instead, the entirety of our action is confined to a single word, or in geometric terms, a highly confined point in visual space.

Although space plays an important role in attentional allocation, as we have already seen through numerous examples in this chapter, our experience with Figure 9.3 shows that there is another important phenomenon,

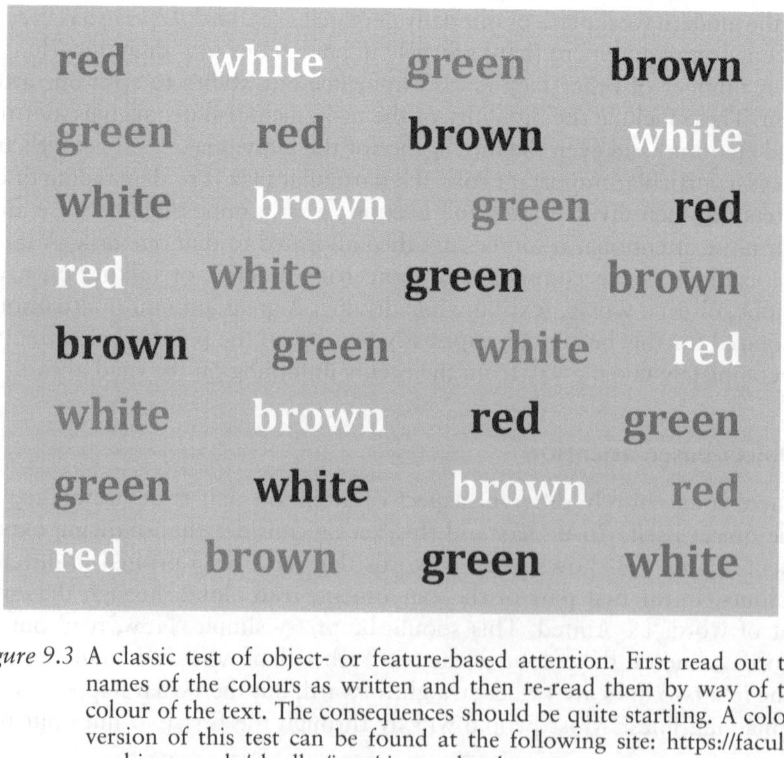

Figure 9.3 A classic test of object- or feature-based attention. First read out the names of the colours as written and then re-read them by way of the colour of the text. The consequences should be quite startling. A colour version of this test can be found at the following site: https://faculty.washington.edu/chudler/java/timestc.html

known as *object-based attention*. It turns out that our attentional system is also concerned with the specific features of objects, whether their colour, shape, texture, movement etc. and that these characteristics have their own capability to draw attention. There have been many studies to convincingly show that we not only direct our attention in a spatially focused manner, but also to the pertinent features of the object that we are processing at any given time. In other words, when we inspect an object, we not only devote our attention to that particular location in our environment, but also to the intrinsic properties associated with that very object.[16,17]

Attentional effects on perception

Now that we have discussed some fundamental aspects of the human attentional system, I next take up what impact this has on our very perception of reality. The reach of attention covers all sensory systems, though the visual and auditory systems are the most affected and therefore have been studied in far greater detail. The role of attention in the somatosensory (touch) system is also well known through our common experience of ignoring tactile stimuli, and even mild pain, by attending to alternate stimuli. There is an enlightened community of healers who specialise in storytelling to help people battle chronic pain. The idea is to move the mind away from the insidious effects of constant pain by focusing attention on pleasant and compelling stories. This healing approach is becoming practiced in areas as diverse as children's care to palliative settings.

Attentional filtering: early selection

Attention has been thought of as a filtering process—our mind selects things that are important and concentrates on them, while discarding the vast amount of irrelevant stimuli that we are being constantly bombarded with. Psychologists have spent considerable effort on trying to identify how the filtering effect takes place and where along the hierarchy of sensory processing it is effectuated. Several models have been proposed to account for experimental data. We will consider two rather different models, shown in Figure 9.4.

One of the earliest models of the human attentional system proposed that stimuli are filtered at a very early stage, and only those that meet a certain threshold of relevance are passed through.[18] This idea, which is illustrated in the left panel of Figure 9.4, is known as *early selection*. Physical stimuli captured by our sensory systems are filtered at an early pre-attentive stage in order to cope with the vast amount of signals captured by our senses. Attention, whether arising from bottom-up or top-down processing, acts upon the incoming signal stream to select those to be passed on for higher-level perceptual processing and storage into

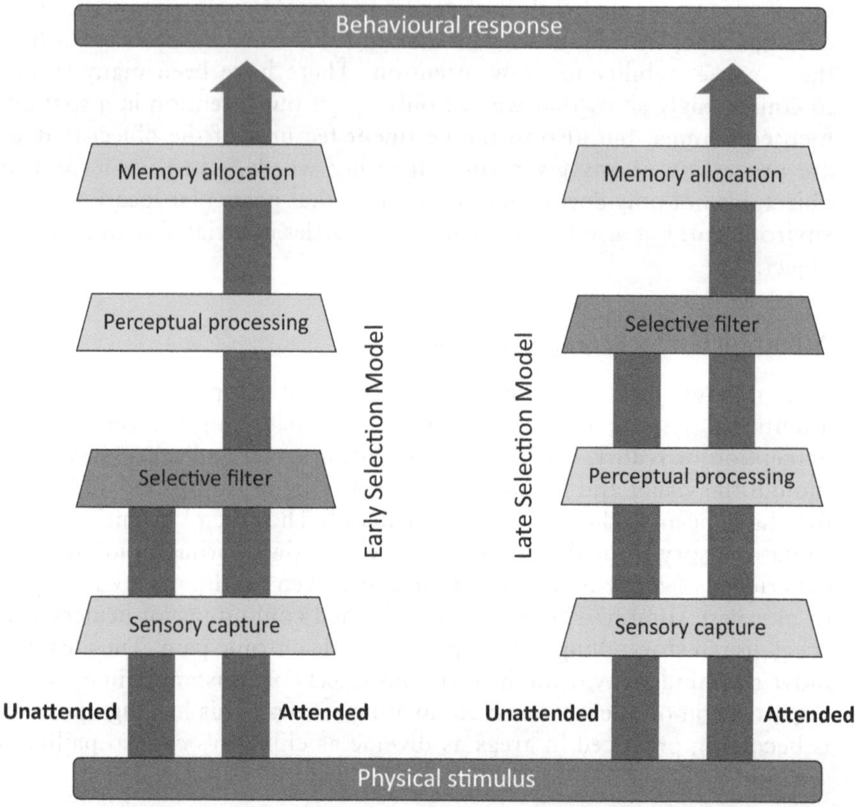

Figure 9.4 Physical stimuli are captured by our sensory systems regardless of whether they are attended to or not. Attended stimuli, however, pass through a selective filter that is located prior to perceptual processing and extraction of meaning in the *early selection* model, and after perceptual processing in the *late selection* model. Those signals that pass through are allocated to memory and may generate a behavioural response.

memory. An attractive feature of this model is that an early filtering process fits in nicely with the idea that the brain would not have evolved to undertake massing information processing only to discard data at a later stage.

Attentional filtering: late selection

The *early selection model*, however, has difficulty in explaining some important experimental findings. One of the key ones is the *cocktail party effect*. We have a striking ability to capture a specific conversation in a crowded room when we are particularly interested in what is being said—perhaps something salacious or personal—even though we may be listening to others in our immediate vicinity. The phenomenon was first described in the 1950s by British psychologist Edward Colin Cherry.

The cocktail party effect shows that unattended stimuli can still be processed with regard to their meaning—for example, the mere mention of our name in the party setting. This finding suggests that all stimuli are initially processed in terms of their meaning or semantics. Then, if the information happens to be relevant, it is passed on to higher stages. This is the construct of the *late selection* model, shown in the right panel of Figure 9.4. Selective filtering, according to this model, takes place after perceptual processing, and only those signals deemed to be relevant at this stage are then allocated to memory.[19] In terms of the cocktail party effect, the *early selection* model would propose that we would not even capture the fact that our name was uttered by someone in that noisy midst, whereas the *late selection* model suggests that all information is perceptually processed and then discarded (or not). Hence, we are able to recognise our name and then allocate our attention only to that conversation.

Perceptual blindness from inattention

We discussed earlier the role that divided attention can have on perception, especially when the demand for a particular task is so great as to capture all or nearly all of our attentional resources. We reviewed a particularly troubling social epidemic in this regard: the use of cell phones while driving. Whether attentional filtering occurs at an early or late stage, the ultimate interplay between memory, perception and our conscious awareness of a sensory signal requires direct attention. Without it, no conscious perception can take place.[20]

Our inability to perceive a stimulus when it is clearly present because of a lack of attention is referred to as *inattentional blindness*. There are many excellent demonstrations of this effect. One of the most forceful is the *invisible gorilla* demo, created by American psychologists Daniel Simons and Christopher Chabris. Subjects are asked to look at a video in which two teams—one wearing black jerseys and the other white—are each passing a basketball only to their fellow teammates. The observer is asked to count how many times one of the teams passes the ball. Nearly half of all observers become so focused on counting the passes that they totally miss the fact that a woman in a gorilla suit enters the court, thumps her chest and walks off. This is an astonishing result and illustrates the powerful effect that attention has on perceptual awareness, as well as the ensuing *blindness* from inattention. A lengthy sequence involving the salient image of a charismatic gorilla-suited individual is filtered out entirely by the attentional system.[21] It is worth trying to find this video on the Internet, though obviously, it will not work on a reader having just read the above passage.

Biological basis of attentional modulation

There has been considerable progress in our understanding of the neural basis of attention over the past three decades.[22,23] Electrophysiological experiments on monkeys have shown that neuronal firing in some areas of the visual cortex is enhanced when there is overt attention to the stimulus. Similarly, if attention is drawn away from the stimulus, then the neuron becomes much less responsive, even though all other characteristics of the stimulus remain the same. In other words, the only parameter in this case that influences the firing rate of the neuron is the level of attention paid to the stimulus. Neuroimaging studies in humans have provided evidence that the physiological underpinnings of attention arise through coordinated neural activity in a network of brain areas that span the frontal and parietal lobes.

Electrophysiological studies

The earliest studies of attentional modulation on neural activity were undertaken with awake, behaving monkeys and largely on neurons in higher visual areas of the brain.[24] A combination of brain imaging and psychophysical studies, in conjunction with electrophysiological experiments, later showed that the effects of attention were not just confined to the higher cortical areas, but in fact affected neurons even at relatively early sites in the visual pathway, such as the primary visual cortex. A particularly tantalising finding was that neurons even earlier in the visual pathways might serve an important gatekeeping function by controlling what information actually flows through them.[25,26]

Neuroimaging studies

An impressive number of brain imaging studies in humans with regard to attention have been carried out using EEG, PET and fMRI techniques. Spatial and object-based attention were the focuses of an influential set of early studies to show the neural structures mediating these processes.[27] The subject is first cued as to which location to attend to. This is similar to the Posner cuing task for allocating spatial attention. The second part of the experiment is similar, except the objective is to study feature-based attention. The cue in this case indicates the direction of a moving stimulus, rather than the appearance of a stimulus in a particular location.

The results of both studies show a remarkable similarity. The visual cue itself created a transient response in early sites of the visual pathways, which is to be expected. A more sustained response, however, was seen in the frontal and parietal cortices of both cortical hemispheres. Brain activity in these regions is believed to be related to the goal-directed attentional requirement of the study because it was not associated with the visual stimulus itself, but to the time period in which the subjects had to direct their attention. In other words, the activity reflects the cognitive operations underlying the attention component of the task. It turns out that the same brain networks are involved in both spatial attention and object-based attention.

The preceding results showed the involvement of a fronto-parietal network for goal-directed attention, whether to spatial or feature-based demands. In this case, the attentional modulation being studied was that of a top-down nature—the subject was made aware of an impending requirement for attentional allocation. This raises the question as to what networks are involved in stimulus-driven, bottom-up attentional recruitment—such as after the sudden appearance of a stimulus. It turns out that in this case, a somewhat different set of brain areas are involved—a more ventral set of structures in the frontal and parietal lobes of the right hemisphere as opposed to the dorsal bilateral set involved in goal-directed attention. This network is involved in directing our attention in an automated manner to a sensory stimulus that has either appeared suddenly or to which we need to suddenly allocate our attentional resources for greater scrutiny.

Perceptual set

As discussed above, attention—or the lack of it—can modulate the perceptual experience to the point where we may not even see something so obvious as a person in a gorilla suit entering our visual field. Similarly, the nature and content of our prior experiences with a stimulus, which is retained in our memory systems, can produce compelling effects on how we perceive that stimulus. The phenomena that have been discussed thus far have led a group of psychologists and cognitive scientists to make a forceful case for a constructivist theory of perception, which stresses the role of knowledge and other cognitive operations in perception.

We now complete our discussion of this issue by reviewing a set of causes that together affect the *perceptual set*—a tendency to experience stimuli based on a set of cognitive factors. There are four general factors that together influence the perceptual set: expectation, motivation, emotion and culture. The role of *expectation* takes place within the context of memory, which is a field vast enough to be outside the scope of this chapter. The remaining factors are briefly taken up in this section. There is a large body of literature on the many fascinating ways in which our perceptual experience can be affected by each of these factors.

Motivation

Let us make an assumption to start our discussion. The reader of this chapter is carefully attending to all that is written here. In other words, attention is focused specifically on the stimulus (visual) to apprehend the material. But why? What is it that is driving the attentional system in this situation? This very question illustrates how an internal cognitive force compels us to behave in certain ways that in turn drives our actions, such as the purposeful allocation of attention to a stimulus or how much effort we are willing to exert toward an outcome. The process that guides our actions in this regard is referred to as *motivation*.

Psychologists have known for decades that a wide range of motivational effects can influence perception. One is the effect of bodily needs such as hunger on how food products are perceived, where certain foods appear to be brighter in colour.[28] The upshot here is that an internal state (hunger) drives the motivation to eat something and in turn makes the food appear brighter or more alluring. Another finding comes from a classic study on how poor and rich children perceive the size of money.[29] When presented with images of coins, children from *poor* families significantly overestimated the size of every coin compared to children from *well-to-do* families. It has been argued that these results show that motivation can affect the perception of size when it comes to money. In other words, the larger coin perception among poor children—or more accurately, the larger perception of their memory of coins—was likely stimulated by a motivational state to perceive something that is more coveted.

A much more recent study has shown how ambiguous figures are perceived based on motivational influences.[30] The perception reported by subjects appears to be driven by what is better in terms of the outcome when they are presented with figures that can be interpreted in one of two ways. The number '13' can be made to look like the letter 'B' if the two numerals are brought close together. Let us say that this stimulus is presented, and if you perceive the letter, then you are given a nice-looking glass of freshly squeezed orange juice to drink, and if you perceive the number, then you are given a glass of disgusting looking green goo ('a gelatinous, chunky, green, foul-smelling, somewhat viscous concoction labeled as an *organic veggie smoothie*').[31] Well, it turns out that subjects consistently perceived the stimulus that would earn them the preferred beverage. In other words, they were motivated to avoid the green goo and therefore, their perceptual experience was accordingly affected. A similar result was found with an ambiguous pictorial stimulus that could be perceived either as a horse or a seal.

There is now ample evidence that our perception of tastes, sounds and sights can be affected by our motivational state, which in turn can be affected by our desire to eat (or drink) something nice and avoid something that appears unpleasant. It remains unclear as to how much of this effect is the result of an automatic outcome that is outside of our conscious will to perceive something. There are many examples where we can consciously make ourselves perceive a stimulus in a particular way. Take, for example, Figure 9.5, a well-known ambiguous stimulus known as the *young girl/old woman* illusion that first appeared in a German postcard in 1888. Looking at this figure in a passive manner creates a stunning effect—an impression of either an old woman looking to the right or a young girl looking away. Only one of these percepts will appear at any given time because the two versions cannot

Figure 9.5 The young girl/old woman illusion. The two possible percepts flip back and while looking at this image. Adapted from Boring, 1930.[32]

coexist. At some point, the impression flips from one version to the other and continues to do so while gazing at the image. What is also clear is that once we have seen both versions, it is possible to consciously force ourselves to perceive only one version of this image for lengthy periods. In other words, we can motivate ourselves to seeing just the young woman by mental effort.

Emotion

It is often believed that people who are upbeat and have a positive view of life have enhanced awareness because their positive emotions encourage greater receptivity to information, which in turn leads to greater creativity. Negative emotions such as sadness, fear and anger, on the other hand, cause an individual to be more constricted in their sensory outreach. This thinking leads to the idea that negative emotions make a person focus on specific details, whereas positive emotions provide a broader perspective. A general validation of this idea has been provided by neuroimaging studies, with the finding that positive emotions produced expanded activity to peripheral stimuli in various brain areas, whereas negative emotions tended to decrease the encoding of peripheral information.[33]

Although it appears that emotion can affect data gathering of sensory signals, does it necessarily mean that emotion can directly affect the perceptual experience? There is now an impressive body of evidence that this is indeed the case. Emotional states can affect low-level visual processes such as contrast sensitivity, where discrimination performance has been shown to improve after subjects are exposed to a fearful face.[34] It therefore appears that certain emotional states can actually enhance perception by improving performance on stimuli that tap into early visual mechanisms. The opposite effect has also been claimed. A phenomenon known as *perceptual defense* proposes that we protect ourselves from offensive stimuli such as painful images or vulgar language by preventing the sensory signals from being processed as quickly as ordinary stimuli. This theory has been criticised on various grounds, however. For example, how can we defend ourselves from a stimulus that we have yet to perceive?

It has also been reported that the vividness of perception is linked to the emotional state.[34] Visual images that are emotionally uplifting are seen more vividly than those that are emotionally disturbing. For example, the amount of noise introduced into a picture is perceived to be less when a positive image is used, with the accompanying result that the underlying picture is seen with greater clarity. It also turns out that these psychophysical findings dovetail nicely with brain imaging studies that show the amygdala—a brain area where emotional processing takes place—to be more active with perceptual vividness. The enhanced activation of the amygdala in turn appears to influence the activity of the visual cortex. It was also found that the greater the vividness of the perception, the more vividly it is retained in memory.[35] And so we come full circle: Our emotional state can determine how vividly we perceive a stimulus, which in turn is the result of enhanced activation in brain areas dedicated to emotion and visual processing, and this in turn creates more vivid and longer-lasting memories.

Culture

We conclude our discussion of cognitive influences on the perceptual set with some fascinating accounts that have appeared from studies on how cultural differences can affect perception. It has long been believed that perception can be influenced by innate bias, tradition and environmental factors—all of which are hallmarks of cultural predisposition. It would therefore not be surprising that the collective aspects of belief, behaviour and environment that define a particular culture should influence the perceptual experience in humans.

An intriguing example of how an environment peculiar to a particular culture can affect visual perception comes from an anecdotal account of the BaMbuti pygmies, a tribe that is indigenous to the dense rainforests of the Congo.[36] Given the nature of their environment, with its heavily forested texture and lack of open spaces, it can be assumed that the BaMbuti people might be impaired at making judgments of absolute depth. This indeed seems to be the case. When a BaMbuti youth was shown wild animals grazing at a distance, something he had not previously experienced, he mistook them for small insects rather than say a herd of buffaloes. In fact, he thought it was magic that the animals would appear larger as he got closer to them. The effective absence of open spaces in the BaMbuti environment meant that his faculty to perceive absolute depth was compromised, and with it, so was the ability to experience size constancy.

One of the definitive works on the influence of culture on perception was carried out by Marshall Segall in the 1960s. It was not clear at the time as to how much, or even whether, cultural differences had an influence on vision. If so, it was reasoned that people from different cultures might perceive certain geometric illusions differently as a result of differences in their environmental makeup. Segall discovered that this was indeed the case after investigating seventeen different cultures. He found that people from American urban centres were more susceptible to certain geometric and optical illusions compared to people of various non-Western cultures.[37]

These results were explained in terms of a so-called *carpentered-world* hypothesis. The presence of a visual world that is rich in oriented stimuli, parallel lines, square corners and other similar structures creates a physical environment in Western society that engages its inhabitants more with perspective cues. Consequently, we are more susceptible to visual illusions based on geometric features compared to people who inhabit a more homogenous environment. The pioneering work by Segall laid the foundation for what is now a widely accepted view: that cultural difference in our environmental upbringing, our attitudes and our beliefs can affect the way we perceive the world. There had been a running debate between an anthropologist (Melville Herskovits), who believed that cultural differences had a strong impact on perception, and a psychologist (Donald Campbell), who believed that all cultures were homogeneous in terms of their perceptual experiences. To resolve this argument, they made their student (Marshall Segall) undertake a six-year study to obtain data from 15 different societies around the world, resulting in his now-classic work on the influence of culture on perception.

Conclusion

The fascinating aspects of our attentional system create an important outcome that was discussed earlier—whether attentional filtering occurs at an early or late stage, the ultimate interplay between memory, perception and our conscious awareness of a sensory signal requires direct attention. Without it, no conscious perception can take place. And without a conscious apprehension of reality, isn't that *something* really a *nothing*?

We do have an appreciation for how Einstein and Tagore would answer that point and argue it from an existential perspective. However, if we are surrounded by real objects and events in a very physical world and yet do not comprehend their existence because of our mental state, then what can be said of that reality? Or if our cultural predisposition is such that we do not perceive the true nature of an object, then how can that distortion be attributable in any way to the physicality of the object itself?

The physicists' point of view argued by Einstein is obvious and actually less interesting because there can be no argument on the physical nature of reality—the atoms are there and together they define a thing that has dimension, creates gravitational force (no matter how small), interacts with electromagnetic radiation in predictable ways and so on. If, however, *reality* is defined purely as a manifestation of the human mind, then Tagore's thesis must be extended to the phenomena that affect the human experience of that reality by way of the cognitive operations that can alter perception. That gorilla was very real, but not to the person who failed to see it.

Notes

1 Gibson, J. J., *The Ecological Approach to Visual Perception* (Boston, MA: Houghton Mifflin, 1979).
2 Gregory, R. L., 'Knowledge in perception and illusion', *Philosophical Transactions of the Royal Society of London B* 352 (1997), pp. 1121–28.
3 Wright, R. D. and Ward, L. M., *Orienting of Attention* (New York: Oxford University Press, 2008).

4 Enns, J. T., *The Thinking Eye, the Seeing Brain. Explorations in Visual Cognition* (New York: W. W. Norton, 2004).
5 Posner, M. I. and Petersen, S. E., 'The attention system of the human brain', *Annual Review of Neuroscience* 13 (1990), pp. 25–42.
6 Schneider, W. and Shiffrin, R. M., 'Controlled and automatic human information processing: 1. Detection, search, and attention', *Psychological Review* 84 (1977), pp. 1–66.
7 Gilbert, C. D. and Li, W., 'Top-down influences on visual processing', *Nature Reviews Neuroscience* 14 (2013), pp. 350–63.
8 Connor, C. E., Egeth, H. E. and Yantis, S., 'Visual attention: Bottom-up versus top-down', *Current Biology* 14 (2004), pp. 850–52.
9 Caparos, S. and Linnell, K. J., 'The spatial focus of attention is controlled at perceptual and cognitive levels', *Journal of Experimental Psychology* 36 (2010), pp. 1080–107.
10 Eriksen, C. W. and St. James, J. D., 'Visual attention within and around the field of focal attention: A zoom lens model', *Perception and Psychophysics* 40 (1986), pp. 225–40.
11 LaBerge, D., 'Spatial extent of attention to letters and words', *Journal of Experimental Psychology* 9 (1983), pp. 371–79.
12 Awh, E. and Pashler, H., 'Evidence for split attentional foci', *Journal of Experimental Psychology* 26 (2000), pp. 334–46.
13 Castiello, U. and Umilta, C., 'Splitting focal attention', *Journal of Experimental Psychology* 18 (1992), pp. 837–48.
14 Sternberg, R. J., *Cognitive Psychology*, 6th edn. (New York: Wadsworth, 2012).
15 Phelps, E. A., Ling, S. and Carrasco, M., 'Emotion facilitates perception and potentiates the perceptual benefits of attention', *Psychological Science* 17 (2006), pp. 292–99.
16 Wühr, P. and Waszak, F., 'Object-based attentional selection can modulate the Stroop effect', *Memory & Cognition* 31 (2003), pp. 983–94.
17 Duncan, J., 'Selective attention and the organization of visual information', *Journal of Experimental Psychology: General* 113 (1984), pp. 501–17.
18 Broadbent, D., *Perception and Communication* (London: Pergamon Press, 1958).
19 Deutsch, J. A. and Deutsch, D., 'Attention: Some theoretical considerations', *Psychological Review* 70 (1963), pp. 80–90.
20 Mack, A. and Rock, I., *Inattentional Blindness* (Cambridge, MA: MIT Press, 1998).
21 Chabris, C. and Simons, D., *The Invisible Gorilla* (New York: Harmony Press, 2011).
22 Nobre, K. and Kastner, S., *The Oxford Handbook of Attention* (New York: Oxford University Press, 2013).
23 Mangun, G. R., *The Neuroscience of Attention* (New York: Oxford University Press, 2012).
24 Desimone, R. and Duncan, J., 'Neural mechanisms of selective visual attention', *Annual Review of Neuroscience* 18 (1995), pp. 193–222.
25 McAlonan, K., Cavanaugh, J. and Wurtz, R. H., 'Guarding the gateway to cortex with attention in visual thalamus', *Nature* 456 (2008), pp. 391–94.
26 O'Connor, D. H., Fukui, M. M., Pinsk, M. A. and Kastner, S., 'Attention modulates responses in the human lateral geniculate nucleus', *Nature Neuroscience* 5 (2002), pp. 1203–9.
27 Corbetta, M. and Shulman, G. L., 'Control of goal-directed and stimulus-driven attention in the brain', *Nature Reviews Neuroscience* 3 (2002), pp. 201–15.

28 McClelland, D. C. and Atkinson, J. W., 'The projective expression of needs: I. The effect of different intensities of the hunger drive on perception', *The Journal of Psychology: Interdisciplinary and Applied* 25 (1948), pp. 205–22.
29 Bruner, J. S. and Goodman, C. C., 'Value and need as organizing factors in perception', *Journal of Abnormal Social Psychology* 42 (1947), pp. 33–44.
30 Balcetis, E. and Dunning, D., 'See what you want to see: Motivational influences on visual perception', *Journal of Personality and Social Psychology* 91 (2006), pp. 612–25.
31 Ibid.
32 Boring, E. G., 'A new ambiguous figure', *American Journal of Psychology* 42 (1930), pp. 444–45.
33 Schmitz, T. W., De Rosa, E. and Anderson, A. K., 'Opposing influences of affective sate valence on visual cortical encoding', *Journal of Neuroscience* 29 (2009), pp. 7199–207.
34 Phelps, E. A., Ling, S. and Carrasco, M, 'Emotion facilitates perception and potentiates the perceptual benefits of attention', *Psychological Science* 17 (2006), pp. 292–99.
35 Todd, R. M., Talmi, D., Schmitz, T. W., Susskind, J. and Anderson, A. K., 'Psychophysical and neural evidence for emotion-enhanced perceptual vividness', *Journal of Neuroscience* 15 (2012), pp. 11201–12.
36 Turnbull, C. M., 'Some observations regarding the experiences and behavior of the BaMbuti Pygmies', *American Journal of Psychology* 74 (1962), pp. 304–8.
37 Segall, M. H., Campbell, D. T. and Herskovits, M. J., 'Cultural differences in the perception of geometric illusions', *Science* 139 (1963), pp. 769–71.

10 Embodied cognition and the constructivist view of reality

Partha Ghose

Einstein's realism

Reality is a word that carries many meanings and subtlely nuanced concepts. In order to have some reliable understanding of what Einstein and Tagore could have meant when they talked about reality, it is necessary to glean their conceptions from a broad survey of their writings and utterances over many years. One conception of 'reality' is that it exists 'out there' quite independent of us and our knowledge of it. In his *Autobiographical Notes*, Einstein wrote, 'Out yonder there was this huge world, which exists independently of us human beings and which stands before us like a great, eternal riddle, at least partially accessible to our inspection'.[1] He also expressed the view that the human mind tries to grasp this external reality with the help of abstract concepts which are often far removed from our direct experience of facts, concepts which it constructs freely, i.e. in a manner that is *logically independent* of the experiences. This is how he expressed his view:

> The prejudice . . . consists in the faith that facts by themselves can and should yield scientific knowledge without free conceptual construction. Such a misconception is possible only because one does not easily become aware of the free choice of such concepts, which, through verification and long usage, appear to be immediately connected with the empirical world.[2]
>
> A theory can be tested by experience, but there is no way from experience to the setting up of a theory.[3]

He also defined what is to be understood when one talks of physical reality in physics:

> Physics is an attempt conceptually to grasp reality as it is *thought independently of its being observed.* In this sense one speaks of 'physical reality'.[4]

Thus, according to Einstein, physical reality is something that is grasped by a free choice of concepts, i. e. by human *thought* without in any way influencing it. All this makes it clear that Einstein's realism is characterised by (i) an external world independent of the human entity, and (ii) the ability of human beings to grasp this reality as it is, independent of observations, at least partially, through a speculative choice of concepts.

In this context, a conversation Heisenberg had with Einstein after a lecture in Berlin in 1926, just prior to his discovery of the uncertainty principle, is revealing. Heisenberg tells Einstein,

> 'We cannot observe electron orbits inside the atom . . . Now, since a good theory must be based on directly observable magnitudes, I thought it more fitting to restrict myself to these, treating them, as it were, as representatives of the electron orbits.'
>
> 'But you don't seriously believe,' Einstein protested, 'that none but observable magnitudes must go into a physical theory?'
>
> 'Isn't that precisely what you have done with relativity?' I asked in some surprise. . . .
>
> 'Possibly I did use this kind of reasoning,' Einstein admitted, 'but it is nonsense all the same . . . In reality the very opposite happens. It is the theory which decides what we can observe.'[5]

This led Heisenberg to critically examine his new theory to see whether it permitted electron orbits to be observed, and he made his famous discovery of the uncertainty principle, which implied it did not.

However, the close correspondence of speculative thought with physical reality remained an eternal puzzle to Einstein. He wrote,

> One may say 'the eternal mystery of the world is its comprehensibility.' It is one of the great realisations of Immanuel Kant that the setting up of a real external world would be senseless without this comprehensibility.
>
> In speaking here concerning 'comprehensibility', the expression is used in its most modest sense. It implies: the production of some sort of order among sense impressions, this order being produced by the creation of general concepts, relations between these concepts, and by relations between the concepts and sense experience, these relations being determined in any possible manner. It is in this sense that the world of our sense experiences is comprehensible. The fact that it is comprehensible is a miracle.[6]

This miracle would not be a miracle without a deep-seated Cartesian mind-body dualism in Einstein's mind, a dualism which Tagore did not share with him.

As far as causality is concerned, which is a part of reality, Einstein's view is clearly expressed in a letter he wrote to Max Born in November 1944:

> You believe in God playing dice and I in perfect laws in the world of things existing as real objects, which I try to grasp in a wildly speculative

way. I firmly *believe*, but I hope that someone will discover a more realistic way, or rather a more tangible basis than it has been my lot to find. Even the great initial success of the quantum theory does not make me believe in the fundamental dice game, although I am well aware that some of our younger colleagues interpret this as a consequence of senility.[7]

He reiterated this view in his conversation with Rabindranath Tagore on 19th August, 1930:

I believe that whatever we do or live for has its causality; it is good, however, that we cannot look through it.[8]

In 1931, a year later, Einstein contributed a piece titled 'About Free Will' in *The Golden Book of Tagore* published from Calcutta, in which he wrote,

But should the lawfulness of events, such as unveils itself more or less clearly in inorganic nature, cease to function in front of the activities in our brain?

Leaving aside the inconsistency of such a view, the influence of alcohol and other sharply controllable factors on our thoughts, feelings and activities should show very distinctly that determinism does not stop before the majesty of our human will.[9]

Despite all this, Einstein's realism strangely encompassed the presence of an omniscient intelligence. For example, on 24th April, 1929, in response to a cablegram from New York's Rabbi Herbert Goldstein asking, 'Do you believe in God?', he wrote,

I believe in Spinoza's God who reveals himself in the orderly harmony of what exists, not in a God who concerns himself with the fate and the doings of mankind.[10]

He also believed in

the profoundest reason and the most radiant beauty, which only in their most primitive forms are accessible to our minds.[11]

When Tagore said to him, '. . . if there be some truth which has no sensuous or rational relation to human mind, it will ever remain as nothing so long as we remain human beings', he quipped, 'Then I am more religious than you are!'[12]

Einstein's views are clearly different from naïve realism, i.e. the belief that science deals with 'facts' proven once and for all, facts that are independent of our experiences and our 'ideas' about them. The opposite view is that 'reality' is dependent on the mind. It is important to remember, however, that this latter view covers two quite distinct strands, one that is purely idealistic in the sense of Berkeley, and the other holding that 'reality' is constructed by the human mind from sense data provided by an external world. It is this latter view that both Einstein and Tagore seem to have held in common, and yet their conversation seems to indicate that they differed radically. To understand what could have caused this difference (apart from the obvious difficulties caused by the presence of a translator), one needs to examine closely Tagore's view of realism, which he, as a quintessential poet, never expressed systematically.

Tagore's views on reality and causality

Tagore's view on reality was strongly epistemological in character. He was by no means an orthodox idealist, as he is often made out to be. The prevalent perception about him in the West and in India, too, is that he was a mystical Indian sage-poet of some sort. Amartya Sen calls this an 'exoticist' view that was responsible for creating a somewhat exaggerated view of India. That was partly responsible for later disenchantment with him in the West and his near oblivion there.[13] In India, too, he has been revered more as a sage and poet, dramatist, music composer and painter than as an analytical thinker with a strong and abiding interest in science and education and India's political and economic independence and development. The physicist S. N. Bose, who was close to Tagore, had this to say about him:

> Had Rabindranath been concerned solely with his own poetry, he would not have built Santiniketan, nor would he have bothered to go out to the countryside to encourage development of agriculture and cottage industries. He aspired to dedicate all of human knowledge first to the Creator of Life with all reverence and then to put this knowledge into the hands of mankind for them to put to use, with good sense.[14]

In his book *Personality*, Tagore writes,

> Science guides man's rebellion against Nature's rule. . . . Science has a materialistic appearance, because she is engaged in breaking the prison of matter and working in the rubbish heap of the ruins. At the invasion of a new country plunder becomes the rule of the day. But when the country is conquered, things become different, and those who robbed act as policemen to restore peace and security. Science is at the beginning of the invasion of the material world and there goes on a furious scramble for plunder. Often things look hideously materialistic, and shamelessly belie man's own nature. But the day will come when some of the great powers of nature will be at the beck and call of every individual, and at least the prime necessaries of life will be supplied to

> all with very little care and cost. To live will be as easy to man as to breathe, and his spirit will be free to create his own world.[15]

In the Foreword to *Visvaparichaya*, the book he wrote on basic science in 1937 and dedicated to Satyendranath Bose, he writes,

> The universe has hidden its micro-self, reduced its macro-self or shelved it out of sight behind the curtain. It has dressed itself up and revealed itself to us in a form that man can perceive within the structure of his simple power. But man is anything but simple. Man is the only creature that has suspected its own simple perception, opposed it and has been delighted to defeat it. To transcend the limits of simple perception man has brought near what was distant, made the invisible visible, and has given expression to what is hard to understand. He is ever trying to probe into the unmanifest world that lies behind the manifest world in order to unravel the fundamental mysteries of the universe. The majority of people in this world do not have the opportunity or power to participate in the endeavour that has made this possible. Yet, those who have been deprived of the power and gift of this endeavour have remained secluded and ignored in the outskirts of the modern world.[16]

During his conversation with Einstein in Kaputh, Tagore emphasised that reality as portrayed by the new theories of nature was 'an appearance, that is to say, what appears to be true to the human mind and therefore is human, and may be called *māyā* or illusion'.[17] If the new scientific view is *māyā* or an appearance, what is the reality that Tagore had in mind that lay beyond the veil of scientific *māyāvāda*? Although Tagore himself never spelt it out, his view of reality seems to have been strongly influenced by a mystic vision he had one morning at the age of seventeen, culminating in the memorable poem 'The Awakening of the Waterfall'. He had a similar vision later in Santiniketan. He wrote, 'I began to see the world around me through the eyes of my soul'.[18] What he meant by seeing the world 'through the eyes of my soul' is, in my view, central to understanding Tagore's view of reality vis-à-vis that of Einstein. In his conversation with Einstein, he used terms like 'Universal Man', 'super-personal Man' and 'infinite personality', terms that Einstein and many others must have found obscure and smacking of pure subjectivism. He writes,

> We live and move and have our being in the *universal man*. A human infant is born into this material universe and into the *universe of man* at the same time. This latter is a world of ideas and institutions, of stored knowledge and trained habits. It has been built by strenuous endeavours of ages, by martyrdoms of heroic men. Its strata are deposits of the renunciations of countless individuals in all ages and countries. It

> has its good and evil elements,—the inequalities of its surface and its temperature making the flow of life full of surprises.[19]

He went on to say in the same book:

> We have seen that the physical life had its gradual expansion into the mental. The mind of animals is fully engrossed in the search for and knowledge of the immediate necessaries of life. In man's case these objects were more varied and therefore a greater mind-power was requisite. Thus we became aware that our world of present needs is one with a world that infinitely transcends our present needs. We came to know that this world not only provides us with food, but with thoughts in a greater measure; *that there is a subtle relationship of all things with our mind.*[20]
>
> The *consciousness of relationship* dawned in us with our physical existence, where there was separation and meeting between our individual life and the universal world of things; it took on a deeper hue in our mental life, where there was *a separation* and *continual reunion between our individual mind and the universal world of reason*; it widened where there was a separation and combination between the individual will and *the universal world of human personalities*; it came to its ultimate meaning where there was *the separation and harmony between the individual One in us and the universal One in infinity.*[21]
>
> *The personality, which is the conscious principle of oneness, the centre of the relationships, is the reality,—therefore the ultimate object of attainment.* I must emphasize this fact, that *this world is a real world only in its relation to a central personality.* When that centre is taken away, then it falls to pieces, becomes a heap of abstractions, matter and force, logical symbols and even those—the thinnest semblances of reality,—would vanish into absolute nothingness, *if the logical person in the centre, to whom they are related in some harmony of reason, were nowhere.*[22]

These passages show that, although he did not spell them out to Einstein, Tagore had precise notions of 'Universal Man' and 'personality' in his mind when he was talking to Einstein. He realised later that he was misunderstood by Einstein. In this context, the use of the capital letter in the word 'Man' is significant—it is to emphasise its independence and transcendence of the individual 'man' and his 'personality'. His writings before he met Einstein did not have it. But the version of the first conversation with Einstein approved by him and published in *The Religion of Man*[23] has it.

Tagore's position is very close to the Indian concept of *achintya bhedabheda* or 'inconceivable oneness and difference', which avoids the extremes of monism and dualism by interpreting the material and spiritual potencies of the Supreme Being as being simultaneously one with and different from

It. In this context, I think Tagore's translation of the Sanskrit word *purusha* as 'person' has led to misunderstandings. *Purusha* has a meaning that is quite different from the English word 'person', whose ordinary meaning is 'an individual man or woman'.[24] It is in his book *Personality* that Tagore *defined* the word 'personality' to mean '*the universal world of reason*', '*the universal world of human personalities*', '*the conscious principle of oneness*', '*the logical person in the centre*' to whom everything is '*related in some harmony of reason*', the individual person being inconceivably one with and different from the universal world of reason. Thus, Einstein's eternal 'mystery' of the comprehensibility of this world is akin to Tagore's 'inconceivable' oneness and difference between the *One in us and the universal One in infinity*. To Tagore, the scientific reality of the insentient world and its spiritual reality cognised through 'personality' are *complementary features of a single reality*. He was aware that he had not succeeded in convincing Einstein about this, and he was convinced of the incompleteness of Einstein's position. In his 'My Memories of Einstein',[25] he wrote:

> I could readily see that Einstein believed my Universe was limited by human conception, and he was convinced that there was some Truth which was independent of human mind. I claim that the Individual has relationship with the Divine Man, who is within as well as beyond us. Our religion is not cosmic, but it concerns our personal being which lives and has its ideal of good and evil. Science makes no such distinctions of a moral or aesthetic nature, and deals with one factor of existence. Science has nothing to do with personality. On the other hand Religion has nothing to do with mere knowledge of abstract Truth.
>
> Einstein has often been called a lonely man. In so far as mathematical symbols help to liberate the mind from the trivialities, I suppose he is a lonely man. His is what might be called a transcendental materialism, which reaches the frontiers of metaphysics, where there can be utter detachment from the entanglement of the world of self. To me both Science and Arts are expressions of our spiritual nature where they are above our biological necessities, and possessed of an ultimate value.

This last point about the spiritual nature of human beings was systematically developed by Tagore in his Hibbert Lectures in Oxford in 1930, just before he met Einstein at Kaputh. Tagore wrote about that meeting, 'My mind was full of the matter contained in my Hibbert Lectures which I had just delivered in Oxford, during May, 1930, and was now putting into book shape under the title of "The Religion of Man" '.[26] As is well known, Tagore's *Religion of Man* is no religion of the usual kind; in it, 'Universal Man' with his logic, aesthetics and morality occupies the central position. In Chapter 1 of the book he writes, 'The idea of the humanity of our God, or the divinity of Man the Eternal, is the main subject of this book'.

Tagore's philosophical position has some interesting parallels with that of Alfred North Whitehead, who argued strongly against the 'Misplaced Concreteness' of abstract theoretical science and the 'Vacuous Actuality' of matter as such. Whitehead wrote,

> A complete existence is not a composition of mathematical formulae, mere formulae. It is concrete composition of things illustrating formulae. There is interweaving of qualitative and quantitative elements; for example, when a living body assimilates food, the fact cannot be that one mathematical formula assimilates another mathematical formula. The fact is more than the formula illustrated.[27]

This is so reminiscent of Tagore's words: 'When that centre is taken away, then it falls to pieces, becomes a heap of abstractions, matter and force, logical symbols . . .' quoted above.

Quantum and classical reality

The predominant view is that quantum reality is fundamentally different from classical reality. The uncontrollable effect of a measurement on a quantum system makes all observed phenomena intrinsically probabilistic in nature, and the distinction between the observer and the observed is blurred. This robs quantum theory of classical realism in the sense of observer independence. One must distinguish this observer dependence from that in Einstein's relativity theory, which also makes space-time measurements observer dependent. The relativistic observer dependence is a consequence of the absolute character of the velocity of light in vacuum—nothing happens to the observed systems, only space-time measurements by observers in relative motion differ in a predictable way. This is not the case in quantum theory—the observed system is uncontrollably affected by acts of measurement by external agencies, and the individual outcomes are unpredictable in principle.

It is usually claimed that 'entanglement' and 'nonlocality' are exclusive features of quantum reality that are absent in classical reality. Recent developments have, however, shown that this is not the case—entanglement in the sense of nonseparability (nonfactorisability) can occur in classical polarisation optics as well when the propagation and polarisation modes cannot be separated, as in 'singular optics'. For example, when a laser beam is incident on a specially prepared 'fork' hologram, it can be twisted like a corkscrew around its axis of travel. Because of the twisting, the light is cancelled out on the axis, and when projected on to a flat surface, the beam looks like a ring of light with a dark hole in the centre. Such vortex beams have wave fronts that are non-planar, and hence they have no definite polarisation. In other words, the polarisation modes and the propagation modes, which span two distinct Hilbert spaces, become nonseparable and entangled. The putative nonlocality of quantum mechanics is usually inferred from violations of Bell inequalities by entangled quantum states. Interestingly, Bell-like inequalities can be deduced for separable (i.e. factorisable) classical optical states like plane-polarised light beams also, and violations of these inequalities by nonseparable (i.e. entangled) classical optical states have actually been observed[28]. But there is obviously no question of any spooky action-at-a-distance or nonlocality in these cases. Hence, what can be inferred from observations of Bell inequality violations in quantum phenomena is nonseparability and not necessarily nonlocality. Nonlocality can be inferred only if one postulates strong projective measurements resulting in objective/ontic collapse. However, there are many interpretations of quantum mechanics that do away with this controversial postulate, such as Everett's interpretation and its offspring.[29]

Entanglement in classical optics is a consequence of its Hilbert space structure. A common belief about Hilbert spaces is that it occurs only in quantum mechanics. This is far from the truth. Way back in the 1930s, Koopman[30] and von Neumann[31] had shown how to write a Hilbert space theory of classical mechanics. Recently, this idea has been extended to the electromagnetic field as well. I do not wish to go into the technical details of all this, except to emphasise that Hilbert spaces, entanglement and Bell-like inequality violations are not unique features of quantumness—they occur naturally in classical polarisation optics as well.[32]

Embodied cognition

That brings me to the question of cognition. How does the human mind cognise reality? Twentieth-century psychology underwent a revolution with three major developments that occurred in the period 1949–1960s. First, Shannon[33] developed a method of measuring the information flow through a system. Second, Chomsky[34] showed that the then prevalent paradigm of behaviourism was unable to explain how human beings acquire and understand language. Third, digital computers arrived, offering psychologists the 'computational metaphor' of the mind and the power to simulate the mind. According to Neisser, 'the task of . . . trying to understand human cognition is analogous to that of . . . trying to understand how a computer has been programmed'.[35]

A different approach to cognition was developed in the 1990s. Called 'embodied cognition', it assumes that *concepts and categories* are *actively constructed* from an *observer-independent* environment. The 'computer metaphor' is replaced by the 'coupling metaphor' of the mind, which asserts that the form of embodiment, the environment and action constrain cognitive processes. It emphasises the interplay among mind, body and environment, the primacy of goal-directed action unfolding in real time. Cognition is viewed as active construction based on an organism's embodied, goal-directed actions and sensorimotor representations.

Consider the case of an artist who was specially known for his colourful, abstract paintings and who became colour blind after an accident. As a result, the overall character of his experiences changed dramatically. Everything he saw 'acquired a distasteful, "dirty" look, the whites glaring, yet discoloured and off-white, the blacks cavernous—everything wrong, unnatural, stained, and impure'.[36] Consequently, his dreams became colourless, he found food disgusting and other pleasures like sex impossible. He could not even enjoy music because he could no longer experience musical tones by synesthetically transforming them into plays of colour. He progressively became a 'night person'. He is reported to have said:

> I love the nighttime . . . I often wonder about people who work at night. They never see the sunlight. They prefer it. . . . It's a different world. There's a lot of space—you're not hemmed in by streets, people . . . It's a whole new world. Gradually I am becoming a night person. At one time I felt kindly toward color, very happy about it. In the beginning, I felt very bad, losing it. Now I don't even know it exists—it's not even a phantom.[37]

Varela, Thompson and Rosch remark:

> This description provides rare insight into how our perceived world, which we usually take for granted, is constituted through complex and delicate patterns of sensorimotor activity. Our colored world is brought forth by complex processes of structural coupling. When these processes are altered, some forms of behavior are no longer possible. One's behavior changes as one learns to cope with new conditions and situations. And, as one's actions change, so too does one's sense of the world. If these changes are dramatic enough—as in Mr. I's loss of color—then a different perceived world will be enacted[38].

What is even more surprising is that different animals have different dimensionalities of colour vision. For example, all colours that normal human beings see are suitable combinations of red, green and blue, which are our primary colours. Hence, our colour vision is three-dimensional. This is made use of in colour photography and colour television. Biologists have found that squirrels, rabbits, tree shrews, some fish, possibly cats and some New World monkeys are dichromats (i.e. their world is made of two primary colours), primates and bees are trichromats, goldfish and freshwater turtles are tetrachromats and diurnal birds like the pigeon and the duck are pentachromats. This does not mean that gold fish and fresh water turtles see colours better than us. What this means is that their colour space is basically incommensurable with ours, i.e. there is no way of mapping all the distinctions available in four dimensions onto three dimensions without leaving a remainder. To appreciate this, imagine that we add a temporal dimension to our colour space, which means that each colour we see flickers at different rates. Then it would not suffice to say something is pink—one would have to specify whether it is *rapid pink* or *slow pink*, for example. This shows that it is impossible for us to even envision what the colour experiences of animals with other dimensions of colour space may be. Varela, Thompson and Rosch go on to say:

> It should now be apparent, then, that the vastly different histories of structural coupling of birds, fishes, insects, and primates have enacted or brought forth different perceived worlds of color. Therefore, our perceived world of color should not be considered to be the optimal "solution" to some evolutionarily posed "problem". Our perceived world of color is, rather, a result of one possible and viable phylogenetic pathway among many others realized in the evolutionary history of living beings.[39]

The assumption that at least some forms of cognition are constructive is supported by a growing number of theorists from a variety of disciplines. The theoretical assumption is becoming more widely supported in the cognitive science literature.

Another emergent area of cognitive modelling is 'Quantum Cognition'. It has many advantages over theories based on classical logic and classical probability theory.[40] The main motivation for searching beyond the limits of classical logic and classical probability theory as the basis for modelling cognitive processes comes from empirical evidence such as the 'Guppy effect',[41] which cannot be explained by any classical model.[42] What the proponents of 'quantum cognition' want to exploit to overcome the inadequacies of classical theory is essentially the Hilbert space structure of quantum mechanics that naturally incorporates superposition, interference, incompatibility, contextuality and order, and entanglement. They claim that this structure lies at the origin of specific effects in cognition related to the way in which concepts and their combinations carry and influence their meaning. Take the example of the pair of concepts *Pet* and *Fish*, and their conjunction *Pet-and-Fish* or their disjunction *Pet-or-Fish*. While Guppy is a very typical example of *Pet-Fish*, it is neither a very typical example of *Pet* nor of *Fish*. This kind of psychological phenomenon cannot be explained by classical theory. The claim is that quantum modelling gives a natural and satisfactory account of such psychological/cognitive processes and reduces to classical modelling in special circumstances. Two other examples where classical modelling does not suffice are: (i) ambiguous states of persons, such as being neither happy nor unhappy; (ii) the answer to two incompatible questions, such as whether Linda is both a feminist and a bank teller, turns out to be inconsistent with the predictions of classical probability theory.

Although no explicit use is made of quantum mechanics as such in the 'quantum cognition' approach, except its Hilbert space structure, the fundamental question that is left unaddressed is: How does any quantumness survive in macroscopic systems like the brain, which is thermodynamically open? It is well known in quantum mechanics that pure quantum states are fragile and subject to rapid *decoherence*, which quenches their quantum properties.[43,44] I only wish to point out that there is no need to invoke quantum mechanics at all in order to use Hilbert spaces to model cognitive processes, because, as we have seen, paradigmatic classical theories such as Maxwell's electrodynamics also have this structure.

It appears that Einstein and Tagore had basically different notions of cognition, Einstein believing in passive retrieval and Tagore in active construction. In 1937, Tagore wrote a poem that appears to me to be a poetic rebuttal of Einstein's position, a poem with a poetic vision of colour, light, sound and the perception of beauty as 'embodied' in Man:

> It was from my sentience that the emerald
> derived its green, and the ruby its red.
> I turned my eyes upon the sky and there
> was light in the East and to the West.
> I looked at the rose and said: 'You are
> beautiful' and the rose gained its beauty.
> You may say that this is metaphysics and not
> poetry. My answer is that it is the Truth
> and therefore poesy.
> You may call it my vanity but it is so on behalf
> of all men. It is on the canvas of man's vanity
> that the creator displays his art of creation.
> With every breath inhaled or exhaled the
> ascetic strives for negation: for him no emerald,
> no ruby, no light, no rose, neither you nor I.
> On the other hand, He who is beyond all limits
> is seeking by His own will to achieve the limit of man,
> and it is Him we call I.
> There was a confluence of darkness and
> light in the assertion of the Ego, and shape
> appeared and emotions found form.
> As if by magic, negation became a yes, in
> colour and form, in sorrow and joy.
> Do not call this metaphysics.
> My mind is full of joy as I sit in the creative
> court of the cosmic—I, with brush in hand
> and bowls of paint.
> The learned say: The ancient moon with its
> heartless and cynical smile creeps closer to the
> ribs of the world like a messenger of death.
> Some day it will heave a vast tug to the oceans
> and hills and this world of matter and form will
> give place to an Amorphous Void, that will
> swallow all debts and credits of days and nights.

Human action will lose even the pretence of
immortality, and his puny history will be
swallowed in the blackness of eternal night.
I say that Man's departing eyes will wrench the
very colour from day, his fading mind will draw
away the throb of all emotions.
Vibrant energy may still walk from sky to sky,
but nowhere will there be Light.
The musicians will make empty gestures, but
there will be neither instruments nor music.
Bereft of poetry the Maker, engaged in the
calculations of existence devoid of personality,
will sit, desolate and alone, in a sky devoid of colour.
Nowhere in this universe, not in its farthest
reaches, not in spaces beyond eternity's sway
in worlds upon worlds, will these words
resound, 'You are beautiful', and 'I love'.
Will the Maker then, once again, sit in
meditation age upon age, and chant in prayer
'Speak, speak, speak again, say . . ." You are
beautiful". Repeat . . . "I love".[45]

Notes

1 Einstein, A., 'Autobiographical notes' in *Albert Einstein: Philosopher-Scientist*, A. P. Schilpp (ed.) (New York: Biography of Living Philosophers, Harper Torchbooks, 1949), p. 5.
2 Ibid., p. 49.
3 Ibid., p. 89.
4 Ibid., p. 81; italics added.
5 Heisenberg, W., *Physics and Beyond*. Translated by Arnold J. Pomerans (New York: Harper, 1971), p. 63.
6 Einstein, A., 'Physics and reality', *Journal of the Franklin Institute*, March 1936, p. 351.
7 Born, I., *The Born-Einstein Letters: Correspondence between Albert Einstein and Max and Hedwig Born from 1916 to 1955* (London and Basingstokes: Macmillan Press Ltd 1971), p. 149.
8 Appendix 2 in this volume; also in Ghose, P. (ed.), *The Poet and the Scientist* (Kolkata: Visva-Bharati, 2014), p. 87.
9 Home, D. and Robinson, A., 'Einstein and Tagore: Man, Nature and Mysticism' in *Journal of Consciousness Studies* 2(2) (1995), pp. 167–79.
10 Clark, R. W., *Einstein: The Life and Times* (New York: Avon Books 2001).
11 Pais, A., *Einstein Lived Here* (Oxford: Clarendon Press; New York: Oxford University Press 1994), p. 118.
12 Appendix 1 in this volume.
13 Sen, A., *The Argumentative Indian* (London: Penguin Books, 2005), chapter 7.
14 Bose, S. N. in *S. N. Bose: The Man and His Work*, Part II, C. K. Majumdar, P. Ghose.
15 Tagore, R., *Personality* (Bombay and Calcutta, India: Macmillan, 1917 First Edition), pp. 90–1.
16 Tagore, R., *Visvaparichaya* (Calcutta: Visva-Bharati, 1937), author's translation.

17 Appendix 1 in this volume
18 Tagore, *Personality*, p. 133.
19 Ibid., pp. 80–1.
20 Ibid., p. 82, italics added.
21 Ibid., p. 84, italics added.
22 Ibid., p. 98. italics added.
23 Tagore, R., *The Religion of Man: Being the Hibbert Lectures for 1930* (New York: The Macmillan Company, 1931).
24 Collins COBUILD English Language Dictionary
25 Ghose (ed.), *The Poet and the Scientist*, pp. 93–8.
26 Ibid.
27 Whitehead, A. N., *Adventures of Ideas* (London: Cambridge University Press, 1933), p. 187.
28 Ghose, P. and Mukherjee, A., 'Entanglement in classical optics', *Reviews in Theoretical Science* 2 (2014), pp. 1–14.
29 See A. Sudbery. in this volume.
30 Koopman, B. O., *Proceedings of the National Academy of Sciences United States of America* 17 (1931), pp. 315–18.
31 von Neumann, J., *Annals of Mathematics* 33 (1932), pp. 587–642; ibid., pp. 33, 789–91.
32 Ghose, P., 'The unfinished search for wave-particle and classical-quantum harmony', *Journal of Advanced Phyics, Special Issue on Tests of the Foundations of Quantum Mechanics* (American Scientific Publishers, 2015), pp. 236–51 to appear: http:\\arxiv:1502.03208 [quantph] February 2015.
33 Shannon, C. E. and Weaver, W., *The Mathematical Theory of Communication* (Urbana, IL: University of Illinois Press, 1949).
34 Chomsky, N., 'A review of B. F. Skinner's verbal behavior', *Languages* 25 (1959), pp. 26–58.
35 Neisser, U., *Cognitive Psychology* (New York: Classic Edition, Psychology Press, Taylor & Francis, 2014)
36 Varela, F. J., Thompson, E. and Rosch, E., *The Embodied Mind: Cognitive Science and Human Experience* (Cambridge, MA: The MIT Press, 1993), p. 164.
37 Ibid., p. 183.
38 Ibid.
39 Ibid.
40 Aerts, D., 'Quantum structure in cognition', *Journal of Mathematical Psychology* 53 (2009), pp. 314–48: http:\\arxiv:0805.3850v2 [math-ph] and references therein
41 Osherson, D. N. and Smith, E. E., 'On the adequacy of prototype theory as a theory of concepts', *Cognition* 9 (1981), pp. 35–58.
42 Tversky, A. and Kahneman, D., 'Judgments of and by representativeness', in *Judgment Under Uncertainty: Heuristics and Biases*, D. Kahneman, P. Slovic and A. Tversky (eds.) (Cambridge: Cambridge University Press, 1982), pp. 84–100.
43 Zurek, W. H., 'Pointer basis of quantum apparatus: Into what mixture does the wave packet collapse?', *Physical Review D* 24 (1981), pp. 1516–25.
44 Zurek, W. H. and Paz, J.-P., 'Decoherence, chaos, and the second law', *Physical Review Letters* 72 (1994), pp. 2508–11.
45 Kabir, H. (ed.), *Poems of Rabindranath Tagore* (Santiniketan: UBSPD in association with Visva-Bharati 2005), pp. 204–6; originally published as *Hundred and One Poems of Rabindranath Tagore*, 1966.

11 Incompleteness theorems and realities

A tale of three great thinkers

Mihir Kr. Chakraborty

Gödel's incompleteness theorems

In the summer of 1930, at a tea session at Café Reichsrat in Vienna, Kurt Gödel almost casually talked about his incompleteness results to a group of researchers of high standing and having allegiance to the Vienna Circle. They had assembled there to discuss preparations for a visit to Könisberg to attend the ensuing conference on the foundations of mathematics. In this group were present Rudolf Carnap, Herbert Feigl and Fredrich Waismann. At the Könisberg conference, Gödel presented his work done for his PhD thesis completed a year before, where he proved the completeness and independence of the axioms of first-order logic. Only during the evening discussion outside the sessions did Gödel mention his incompleteness theorems once again. John von Neuman was present. He immediately realised that something ultimate as well as extremely devastating had taken place. It is quite likely that he informed his teacher David Hilbert, who delivered a public lecture the next day in which he declared emphatically, 'We must know! [We] shall know'. He was re-uttering his conviction that all mathematical queries were answerable and all mathematical entities were knowable. If one could formulate a mathematical sentence properly, either it or its opposite (negation) would be derivable from a finite set of initial assumptions. In fact, he was never willing to accept Gödel's claim, which would jeopardise his grand hope of establishing the whole of mathematics on a finitistic foundation.

Thus began a new era. In the backdrop, there had been the great debate among Logicism (Russell, Whitehead), Intuitionism (Brouwer) and Formalism (Hilbert) regarding the foundation of mathematics. These are the three distinct grand narratives on mathematics and its foundations, proposed and widely discussed. Gödel's two incompleteness theorems dealt a severe blow to Logicism and Formalism. Intuitionism was already on the back seat. A brief introduction to the main features of the three contending viewpoints is called for at this point.

Logicism views mathematics itself as logic, not merely based on logic. Starting from the axioms of first-order logic that were proposed and developed by Frege and afterwards modified by Whitehead and Russell, and by using the rules of inference, the natural numbers, the primordial mathematical objects, would be constructed at one stage as entities of logic. After that, the transfinite numbers, negatives, fractions, irrationals and complex numbers would be construed. Then the entire structure of mathematics might be built gradually. Thus, all the axioms of mathematics would be derived by using purely logical laws and logical modes of reasoning. However, Logicism a la Russell and Whitehead had to succumb finally due to several reasons, of which one was the observation that, apart from the logical entities, one required the notion of 'class', which cannot be included within 'pure' logic. Besides that, two axioms of Logicism, viz., the axiom of infinity and the axiom of reducibility, generated criticism. Rudolf Carnap represented the logicist school at the Könisberg conference.

Formalism holds that in the ultimate analysis, mathematics has to be based on some 'meaningless', undefined terms and assumed relations among them. Neither these terms nor the assumed relations have necessarily anything to reflect about reality, although sometimes these may be interpreted as real. From this point of view, the terms and relations are only signs, and mathematics turns out to be the game of signs—formal manipulation of symbols. In Könisberg, John von Neumann was the spokesman of formalism.

Intuitionism, on the other hand, proposes to dissociate mathematics from logic—it would be better to say, according to this view, that mathematics is more fundamental than logic. It propounds that an ideal mathematical mind can access mathematical notions. Besides, some of the laws of classical logic, e.g. the law of excluded middle and the principle of double negation, are not acceptable to Intuitionism. Moreover, the existence of a mathematical entity is ensured only by a construction and not by reductio methods. Although Brouwerian principles have been relaxed by later intuitionists, these core doctrines have remained unaltered. During Gödel's time, however, Brouwerian Intutionism had already taken a rear position. Arend Heyting, who differed in many ways from his teacher Brouwer, represented the intuitionistic school in the Könisberg conference. Gödel also took some interest in Intuitionism at one stage and made significant contributions.

To this backdrop were added new questions and controversies by Gödel's incompleteness theorems. After an initial period of scepticism, Gödel's 'proofs' turned out to be correct, although, as it always happens, several other simpler proofs and proofs from other angles appeared over time.

What precisely are these new dimensions? Along with the formalist approach towards mathematics, Hilbert also had a deep-seated belief in founding the entire mathematics on finitary methods. Although this notion was never 'defined', it was clear that some finite set of symbols and finitary rules would constitute such a method. And Hilbert hoped to 'prove' the consistency of mathematics by using such a method. Altogether, this was called 'Hilbert's programme'. This programme got a jolt from Gödel's theorems. Gödel showed that one has to choose between consistency and completeness.

A formal theory, a prototypical finitary method, strong enough to include an elementary fragment of number theory (Peano arithmetic), is incomplete if it is taken to be consistent. This is the first incompleteness theorem. So, assumptions of consistency of mathematics as a formal system lead to its incompleteness, i.e. there shall always remain some mathematical statements (formally, well-formed formulae) that cannot be proved, nor their negations. But precisely its opposite was the hope of David Hilbert, who had declared it several times, including in the aforementioned public lecture by him in Könisberg.

Secondly, even though one is ready to forego completeness of mathematics, are we in a position to 'prove' its consistency? Gödel's second incompleteness theorem says that if the formal system is consistent, its consistency cannot be 'proved' within the system itself. Assuming, according to Hilbert's programme, that the entire mathematics is one mathematical system founded on a finite set of axioms and a set of finitary rules, it follows from the second incompleteness theorem that its consistency cannot be established within the system. Then what remains outside formal mathematics? Can we hope that consistency can be established by a blend of informal, loosely knit argumentation and mathematical intuition? The following quotation from Henry Mehlberg tries to provide an answer to this question, reflecting one shade of attitude:

> What I am driving at is simply the idea that a mathematical theory can be dependably known to be consistent if a reasonable number of appropriately variegated attempts at deriving a contradictory consequence from this theory have been made over an adequate time by competent investigators and if all these attempts have proved unsuccessful. Granted, their failure does not theoretically preclude the possibility that, in a more or less remote future, a lucky investigator may succeed in deriving such consequences from this theory, his theoretical possibility points simply to the fact that our knowledge of the consistency of the theory is not infallible although it is dependable to a very high degree.[1]

So, we are advised to be ready to accept that the consistency of mathematics, say, number theory, as experienced by the community is not infallible. This is a fact hardly palatable not only to the mathematicians, but also to its users (physicists, for instance), as well as citizens of modern states at large. Gödel himself would certainly not accept this understanding of mathematical consistency.

A third point deserves to be emphasised in this context. That is connected with the notion of the truth of mathematical claims or propositions. According to the second incompleteness theorem, the consistency of mathematics implies that it (this consistency) cannot be proved in mathematics when considered as a formal theory. From our long familiarity with numbers, it is reasonable (here I follow Mehlberg) to assume that number theory is indeed consistent. So, by incompleteness theorem, its consistency is not provable. Thus, some mathematical 'truths remain beyond mathematics'—a phenomenon making a sceptic naturally delighted. Gödel himself was far from scepticism in mathematics. He even 'believed' that someday, the continuum hypothesis would be established, though later years proved otherwise. But what Gödel proved in the second incompleteness theorem helps the sceptic in doubting truths obtained through mathematics, and as a consequence, truths in science. On the other hand, mathematicians earning their bread and butter on mathematics would say that the problem lies with the formal system; Gödel's second incompleteness theorem shows only the limitation of formalisation, not of mathematics per se—too much dependency on formalism does not really pay. But, we know that formalism was a necessity of a particular era, at a particular stage of development of mathematics. Various problems arose out of general informal practices continuing over centuries. We know the story of non-Euclidean geometry. The informal arguments for centuries to prove the fifth postulate ended up with circularities several times. Concepts of calculus were extremely amorphous, so much so as to be ridiculed by Bishop Berkeley with the remark that infinitesimals of calculus are 'ghosts of departed quantities'. This experience led mathematicians of the 20th century to take recourse to extreme caution while attempting to repair the damage caused by several paradoxes and antimonies, such as Russell's paradox. In the ultimate analysis, what else can 'caution' mean other than 'rigour', and rigour means looking for some sort of formalism that can be used and inspected by the human brain mechanically, without any intuitive leap.

In mathematics, one cannot depend simply on (mathematical) intuition and informal understanding or argumentation: one needs rigour. Neither is the 'truth' of the basic axioms verifiable by observation nor are all the axioms 'self-evident', as is often professed. In all the clusters of axioms determining various theories in mathematics, there exists at least one which is not 'self-evident', such as the fifth postulate of Euclidean geometry, the Induction hypothesis of Peano Arithmetic, the triangle inequality of metric space axioms or the substitutivity axiom for equality in first-order logic. These are 'mysterious' in a sense and loaded with condensed mathematical matter. Many axioms of set theory are of this kind, i.e. they are not at all self-evident. (*Of course, Gödel maintained exactly the opposite view.*)

Let us scrutinise the derived 'truths', i.e. the theorems. How to ascertain that the claim viz. the angle sum of a triangle is 180 degrees is true? This is a theorem of Euclidean geometry. Repeated actual measurements, i.e. observations or experiments, however accurately measured, are not likely to 'verify' the claim. Certainly, the degree of 'belief' within the users or common people will increase with the increasing number of measurements and greater accuracy, but the 'truth' of the claim will not be 'established'. Hence the slogan: in mathematics, no truth without proof. And proof means derivation by using logical principles from some initial, unproved but accepted claims, the axioms—none of which is physically verifiable—and, as stated before, there exists at least one that is far away from intuitive acceptability or 'self-evidence'.

So, the Hilbertian position stands thus: One has to start with unproven, unverified, not always self-evident, meaningless sentences that in the ultimate analysis are strings of uninterpreted symbols. What Gödel's theorem has demolished is the guarantee that it, the mathematical system that seems consistent so far, would continue to be consistent for all time to come.

Godel's view on mathematics

Gödel himself, however, would not subscribe to the above thesis that mathematics is the game of meaningless signs. He was quite firm in his position that a sign carried meaning which is the actual mathematical object, not the sign. The object is abstract but resides outside the human mind, though this residence is not like that of the physical objects that physics deals with. He, in fact, categorised mathematics as objective and subjective. To him, the body of mathematics that is true and independent of the human intellect is objective, but 'humanly provable mathematical assertions' are subjective. Gödel believed that mathematical truths of the first kind do exist—truths of whole numbers fall under this category. Or set theory, perhaps. Look at what Gödel says of sets:

> . . . despite their remoteness from sense experience, we do have something like a perception also of the objects of set theory as is seen from the fact that the axioms force themselves upon us as being true. I do not see any reason why we should have less confidence in this kind of perception i.e. in mathematical intuition than sense perception.[2]

(*That I differ from Gödel in this point has already been stated.*) But what else? Other than sets and whole numbers, what else is objective mathematics to Gödel is not clear. To him, even Euclidean geometry appears to be subjective.[3] For maintaining belief in objective mathematics, Gödel has remained a Platonic realist in his philosophical position on mathematics. Russell mentioned him as an 'unadulterated Platonist'.[4]

But Gödel suffered too. In spite of his strong adherence to Platonic realism, his rationality did not vouch for what he had been intuiting. His rational self looked for a stronger support to his belief, and that could not be possible. To some Gödel researchers, his leaning towards Husserlian transcendental idealism was due to this failure in him.[5] More specifically speaking, on the one hand, he was a staunch realist, and on the other, he was unable to construct a rational 'proof' for that realism. Rational proof, according to him, was a proof as in mathematics. Thus he wished to see the results of philosophy (of mathematics in this case) as theorems of mathematics, i.e. methods of inference in philosophy as well knit and as demonstrable as in mathematics. Gödel has also been looking for some sort of 'epistemological parity', that is, some (abstract) evidence for the existence of abstract mathematical entities.[6] And it was hard, if not impossible, to be found. His lifelong pursuit to understand the nature of mathematics, a field to which he was himself an outstanding contributor, seems not to have reached a final end. His philosophy of mathematics did not find an adequate shelter in Husserl, too.

Although what Gödel looked for was not quite clear, or better to say, what would have satisfied him could not be stated in clear terms, it is possible to enumerate some of the ideas that he did not definitely entertain. He did not accept that mathematics was a game of meaningless symbols, as propounded by the Hilbertian formalist school. He also opposed the viewpoint that mathematics was only a syntax of language. Nor did he accept that mathematics was a free creation of the human mind, as opined by Carnap et al. His Nachlass was full of writings, finished and unfinished, published and unpublished, for refuting all three of the viewpoints mentioned above.[7] One quotation from the Gibbs lecture that he delivered will make his position and at the same time his sufferings transparent:

> Of course I do not claim that the foregoing considerations amount to a real proof of this view (i.e. Platonism) about the nature of mathematics. The most I could assert would be to have disproved the nominalist view, which considers mathematics to consist solely in syntactic conventions and their consequences. Moreover, I have adduced strong arguments against the more general view that mathematics is our own creation . . . In order to establish platonic realism, these theories would have to be disproved one after the other, and then it would have to be shown that they exhaust all possibilities. I am not in a position to do this now.[8]

My view

In this chapter, I am not presenting actually what Gödel's arguments were, nor critiquing his arguments, for which one can refer to numerous essays, a very good anthology being Jacquett's.[9] There exist critiques of Gödel's views made both from other realist angles as well as from the antirealist standpoints. In the remaining few paragraphs, I intend to present my own viewpoints on some of these issues very briefly and without much elaboration. For a relatively detailed account, I refer to my earlier writings.[10] In my view, a mathematical object may be considered to be an artefact made by a mathematician. In this respect, a mathematician resembles an artisan. The artefact is often rich with so much of aesthetic content or value that it turns into an artistic object or a *RUPA*, and its creator, the mathematician, turns into a mathematical artist. *RUPA* is a Sanskrit word standing for a visual image that is beautiful. But it is used not necessarily for visual images only: mental images are also included and the radiation of beauty constitutes the most important component of the concept. The creation of abstract mathematical objects resembles the creation of an abstract or semi-abstract art by an artist—in this latter case, that is, of art, it becomes more private. In case of mathematics, the initial creation is usually a personal work, and then the community as a whole takes part in its recreation or in giving the object a complete shape. There are plenty of instances of art objects also created by a group or a community. After its creation, the mathematical entity becomes public, as in the case of any other created object. Anybody can establish with the object any kind of relationship compatible with its form and content. Even though various relationships are set up by various takers of the created mathematical object, there usually remains a common thread around which communications among different takers of a particular mathematical object takes place. As Tieszen remarks,

> With quantification, calculation, formulation, idealization and exactness . . . we obtain an inter-subjective argument on methods and results, including repeatability of calculation, experiments and procedure. We obtain a kind of objectivity.[11]

In the making of mathematical objects, language plays a vital role. At the informal or semi-formal stage of the creation, natural language, the language of set theory and logical language take major roles, while at the later or final stage, formal language dominates. Objects are created by axioms, and axioms are formed in a mixture of natural and formal languages. Mathematics is a subject that requires extensive use of symbols, and I do not think it would be possible to dispense with them even in principle. To hold the mathematical mass accumulated in a created mathematical object (a concept), natural language turns out to be inadequate. A special symbol for the object (or concept) is to be invoked. A mathematical text turns into a discourse about such symbols, and at a sufficiently mature stage, the discourse itself turns into symbols—strings of symbols or strings of strings of symbols.

This is about the ontology of mathematical objects like numbers, geometric objects, sets, algebraic and topological structures and so on. New, hitherto unknown properties of mathematical objects are 'discovered' through proofs. So, this is like a discovery in science, but only after the objects are invented or created. Also, proofs are to be considered as mathematical objects. What is understood by a proof in the realm of mathematics is not the same category of items that go under the same name in our day-to-day parlance or even in scientific discourse. This is particularly so when a proof deals with mathematical objects that involve infinite sets and/or the proof procedure itself is infinite, such as by use of induction. It is not the same procedure as discovering the 'truth' of nature as followed in science or discovering the 'truth' behind a social event, say, an incident of crime. When one proves that the number of points in a square is the same as the number of points in one of its sides, the 'truth' established goes against intuition and beyond common understanding. Cantor himself commented in such a context, 'I see it but cannot believe it'.[12] As regards Gödel's proofs of incompleteness theorems, the following remark by Wittgenstein is revealing. He mentioned 'hearing' statements about 'true' and 'false'—that there were true mathematical statements that could not be proved in *Principia Mathematica*. In such cases, his opinion was to avoid the words 'true' and 'false' altogether.[13] In Gödel's view, Wittgenstein did not understand the incompleteness theorems. Current studies show that Wittgenstein was not averse to incompleteness theorems, but did not see eye to eye with Gödel on the notion of mathematical truth. He was not questioning the mathematical proof itself but certain 'philosophical prose' surrounding the proof. In my view also, 'true' stories in mathematics are of a different kind that should not be taken on par with those in natural sciences or society.

My conviction in connection with the activities of humankind centring around mathematics is orthogonal to that of Gödel. But that does not imply that I consider Gödel or traditional mathematicians in general are wrong, and that the mathematics they have produced are inaccurate or inelegant or useless. On the contrary, the domain of classical mathematics has gifted to humanity some of the finest ideas and some of the most effective artefacts for application. Indeed, nobody will be ready to exclude the theory of numbers from mathematics and its consistency from his/her belief base. Classical mathematics is correct because it has played the game correctly as per the rules set by itself, and so far the inconsistencies generated have been mitigated and no new inconsistency is in sight. Gödel's own proofs of completeness of predicate calculus, and the two incompleteness theorems are testimonies of the greatest achievements of human thought. Proofs in set theory are examples of the highest perfection, as perfect as the dance of *Nataraja* of Indian mythology and portrayed by numerous artists. I would only request the readers to keep in mind that non-Euclidean geometries and non-Cantorian set theories are equally elegant and important pieces of human creation. All these alternatives abound in the mathematics scenario nowadays. Conferences are held on the culture dependency of mathematics. Books under radical titles such as *Inconsistent Mathematics*[14] or *Mathematics: the loss of certainty*[15] or *Pluralism in Mathematics*[16] are being published. As opposed to the natural sciences, particularly physics, mathematics has no unique goal, no converging arrows. It has the potentiality to proliferate in different, often opposite directions: This is the essence of mathematics. There is no one mathematics, but many. Before writing the last sentence of this part, let me quote from Sofya Kovalskaya, the first woman appointed to a professorship in mathematics in Europe at Stockholm University in 1889 (*taken from the article Duality by Mai Gehrke, another leading woman mathematician of our time):*

> You are surprised at my working simultaneously in literature and in mathematics. Many people who have never had occasion to learn what

> mathematics is confuse it with arithmetic and consider it a dry and arid science. In actual fact it is the science which demands the utmost imagination. One of the foremost mathematicians of our century says very justly that it is impossible to be a mathematician without also being a poet in spirit. It goes without saying that to understand the truth of this statement one must repudiate the old prejudice by which poets are supposed to fabricate what does not exist, and that imagination is the same as 'making things up'. It seems to me that the poet must see what others do not see, and see more deeply than other people. And the mathematician must do the same.[17]

My *mathematical artist* is completely free to choose her own niche and get absorbed in the creative act by producing new mathematical objects, new *RUPAs* and in unfolding the mysteries hidden in those *RUPAs* which are products of her own imagination. They might get utilised, but that is not her primary responsibility.

Gödel, Einstein and Tagore

At the Institute for Advanced Study in Princeton, Gödel had very few, almost no, friends, but Einstein was an exception. Einstein often remarked that he used to go to the institute to avail the privilege of walking back home with Gödel. During the walk they usually talked intensely. Gödel was deeply interested in Einstein's theory of relativity. He even came out with a new cosmological solution to Einstein's field equations of gravitation.[18] Einstein accepted it.

Did Einstein show any interest in Gödel's incompleteness theorem? It seems that, however strange it might appear, the answer is 'no'. To Gödel-Einstein researchers, particularly in the present era when hermeneutics has entered the field of studies in history of science, this question demands a probe. Gödel's results are devastating in that number theory's consistency is in jeopardy, and as a consequence, that of mathematics as a whole. Physical science being un-detachably dependent on mathematics may naturally feel threatened by Gödel's theorem, as mathematics itself should be. One would not feel at ease in the face of such a danger. Besides, Einstein had already been in great intellectual trouble with the emergence and development of quantum physics. He might not have liked to invite additional trouble on an issue that was rather remote to his main area of concern. We have noticed that in the after years, physicists did not ignore incompleteness theorems altogether. Roger Penrose took resort to a Gödelian argument to establish his ideas on the philosophy of mind, though not on physics per se.[19] But nevertheless, he is not comfortable with two mathematics (logics), one for the subatomic world and another for the other ones. He looks for some alternative mathematics for the grey zone of micro and macro that could serve as a bridge between the two for a seamless passage from one to the other. However, Stephen Hawking is more direct in accepting the influence of Gödel's method in the philosophy of physics. I quote the following paragraph from his talk entitled, 'Gödel and the end of physics':

> What is the relation between Gödel's theorem, and whether we can formulate the theory of the universe, in terms of a finite number of principles? One connection is obvious. According to the positivist philosophy of science, a physical theory is a mathematical model. So, if there are mathematical results that cannot be proved, there are physical problems that cannot be predicted . . . we are not angels who view the universe from the outside. Instead, we and our models are both part of the universe we are describing. Thus a physical theory is self-referencing, like in Gödel's theorem. One might therefore expect it to be either inconsistent, or incomplete. The theories we have so far, are both inconsistent, and incomplete.[20]

Thus, with the loss of certainty in mathematics, loss of certainty in physics also creeps in. This uncertainty is not due to limitation of observation, nor due to the possibility of appearance of new phenomenon hitherto unknown, rendering fallibility to science. This is more deep-rooted, due to uncertainty in the validity of mathematical claims and laws as used in physics.

This utter dependency of physics on mathematics is what Tagore hinted at in his dialogue with Einstein in 1930: 'You have been busy, hunting down with mathematics, the two ancient entities, time and space, while I have been lecturing in this country on the eternal world of man, the universe of reality'.[21] There are at least two versions of the dialogue: one recorded by Dimitri Marianoff in *The New York Times*,[22] and another published in *Modern Review*[23] These do not perfectly match, though it is clear what the central issue was that the two great minds conversed upon (*as if 'two planets engaged in chat' a la Marianoff*). As described by Marianoff, Tagore was a poet with the head of a thinker and Einstein was a thinker with the head of a poet. In this chapter, I am using both of the above-mentioned records and also the record by Sudhir Ghose of the second conversation in Berlin.[24] They differed. While Einstein held that truth lies outside human consciousness (mind), Tagore said, 'This world is a human world—the scientific view of it is also that of a scientific man. Therefore, the world apart from us does not exist; it is a relative world, depending for its reality upon our consciousness'.[25] It seems that Tagore was aware of the developments in quantum theory and ongoing controversies on its foundation. So he was pointing at the new mathematical discoveries that professed that in the realm of infinitesimal atoms, chance had its play, existence was not absolutely predetermined in character. At one stage, Tagore asks straightway, 'Are the elements rebellious, dynamic with individual impulse? And is there a principle in the physical world which dominates them and puts them into an orderly organization?'[26] Einstein's answer was that even if there were chances, there were laws of statistical order; in any case, the law of causality was not done away with. This interpretation means that although there remains uncertainty about the role of individual atoms, the role of the crowd is quite accurately predictable by the statistical laws. It should be noted that Einstein himself was sceptical about the statistical interpretation of the ontology of reality; he believed that God does not play dice. He was arguing only in favour of causality. But two points remain unanswered: First, how is the behaviour of a crowd determined? Is it not determined by the behaviour of individual members of the crowd? And the second, the main point, is whether the phenomenon that uncertainty prevails in the behaviour of individuals while the crowd consisting of these individuals remains predictable resides as a fact outside human consciousness (mind). Einstein said 'yes', and Tagore did not agree. Tagore held that the truth that is described as scientific can only be reached through the process of logic, in other words, by an organ of thought which is human. In 'My Memories of Einstein', Tagore writes a very interesting sentence, 'While logic belongs to man's mind I can conceive of another field of logic for another type of mind other than ours'.[27] It was evident to him that through human reasoning, facts assume a unity of truth that is only possible to a human mind. Thus the truth is humanly constructed, constructed by reason that includes facts (observation by senses) and logic (including mathematics). If the logic differs, the conclusion, the truth, might differ too.

All the three thinkers, Gödel, Einstein and Tagore, had been realists. Gödel, as mentioned before, even believed in the existence of mathematical entities (objective mathematics) outside the human mind, not only in the existence of physical reality. His problem was that he also cherished the hope of having a sort of 'proof' of this belief, but that was not available. Einstein considered mathematics (or logic) as only a tool for explaining physical phenomena. Although he had to liberate himself from Euclidean geometry, he could not accept the new statistical method adopted in quantum mechanics to understand the sub-atomic world. It seems that, to him, other problems of mathematics, namely its consistency or completeness, were not an issue deserving that importance. Tagore's realism is embedded in traditional Indian philosophy. According to him, the absolute truth was Brahman, which could not be conceived by the isolation of the individual mind or described by words, but could be realised only by merging the individual in its infinity.[28] And further, 'what we call truth lies in rational harmony between the subjective and objective aspects of reality, both of which belong to the super-personal man'.[29] Tagore is also a believer of a unique all-pervasive truth existing outside the individual mind but which would be realisable by the 'universal mind'. Without the hypothesis of this 'super-personal man' or 'universal man', this truth, which is unique, would be in a sense non-existent. It is a reality constructed not by individual minds but by the universal mind. Tagore mentioned other possible logics. In fact, the present era has produced several logics, many of them representing actual argumentation procedures adopted by the human mind. It would be irrational to presume that to explain or understand physical reality, only one logic or mathematics would be appropriate. If different alternative logics are accepted, different alternative causes to explain the same physical phenomenon may be arrived at. Neither Tagore, nor Einstein nor Gödel would be ready to accept this kind of pluralistic view of reality.

I would like to end with the statement of a few facts that might be interesting to the readers. The Tagore-Einstein dialogues took place on 14 July, 1930, at Kaputh and on August 19, 1930, in Berlin. Gödel announced privately his incompleteness results at Café Reichsrat to Carnap, Feigl and Waismann on August 26, 1930. What a wonderful summer!

Notes

1 Mehlberg, H., 'The present situation in philosophy of mathematics', *Synthese* 12 (4) (1960), pp. 380–414. Reprinted in Jacquette, D. (ed.) (Malde, Massachusetts: Oxford), *Philosophy of Mathematics, an Anthology* (USA & UK: Blackwell Publishers, 2002).

2 Quoted from Maddy, P., 'Sets and numbers', *Noüs* 15 (4) (1981), pp. 495–511. Reprinted in Jacquette(ed.), *Philosophy of Mathematics*.

3 Feferman, S., Dawson, J. W. Jr., C. Parsons, W. Goldfarb, and R. Solovay (eds.), *Kurt Gödel Collected Works*, Vol. 1, 2, 3, 4 (New York: Oxford University Press, 1995).

4 Dawson, S. (ed.), *Kurt Gödel, the Album* (Mühlberger: Vieweg, 2006).

5 Atten, Van M. and Kennedy, J., 'On the philosophical development of Kurt Gödel', *The Bulletin of Symbolic Logic* 9(4) (December 2003), pp. 425–76.

6 Ibid.

7 Feferman, Dawson, Parsons and Solovay (eds.), *Kurt Gödel Collected Works*.

8 Ibid., Vol. 3, pp. 321–22.

9 Jacquette (ed.), *Philosophy of Mathematics*.

10 Chakraboty, M., *Ganiter Dharapat o Galposalpo*, 4th edn. (Kolkata, India: Nandimukh Samsad, 2011).

Chakraboty, M., *Gödel-er Asampurnata Tatva* (Kolkata, India: Nandimukh Samsad, 2012).

Chakraboty, M., 'Mathematical rupas and their artists', in *History of Science and Philosophy of Science: A Philosophical Perspective of the Logic of Ideas in Science*, Pradip Kumar Sengupta (ed.), PHISPC, Vol. XIII, Part 7 (Delhi: Pearson, 2012), pp. 527–33.

11 Tieszen, R., 'After Gödel: Mechanism, reason and realism in the philosophy of mathematics', *Philosophia Mathematica* 14 (2) (June 2006), pp. 229–54.

12 See Gouvea, Q., Was Cantor surprised: http://www.maa.org/pubs/AMM March11_Cantor.pdf (2012) for an interesting account.

13 Diamond, C. (ed.), *Wittengenstein's Lectures on the Foundation of Mathematics*, Cambridge 1939 (Chicago and London: The University of Chicago Press, 1975). The reference taken from Floyd, J., 'Prose versus proof: Wittgenstein on Gödel, Tarski and truth', *Philosophia Mathematica* 3(9) (2001), pp. 280–307.

14 Mortensen, C., *Inconsistent Mathematics* (Dordrecht: Kluwer Academic Publishers, 1995).

15 Kline, M., *Mathematics, the Loss of Certainty* (Oxford, NY: Oxford University Press, 1980).

16 Michele, F., *Pluralism in Mathematics; A New Position in Philosophy of Mathematics* (Dordrecht: Springer, 2014).
17 Gehrke, M., 'Duality', *Radboud Universiteit Nijmegch* (2009), p. 5.
18 Kurt, G., 'An example of a new type of cosmology solution to Einstein's field equations of gravitation', *Reviews of Modern Physics* 21(3) (1949), pp. 447–50.
19 Penrose, R., *Shadows of the Mind* (Oxford, NY: Oxford University Press, 1994).
20 Franzen, T., *Gödel's Theorem; An Incomplete Guide to Its Use and Abuse* (Massachusetts: A. K. Peters, 2005), pp. 88–9.
21 'Einstein and Tagore plumb the truth', *New York Times*, 10 August 1939.
22 Ibid.
23 Tagore, Rabindranath, 'The nature of reality', *Modern Review*, January 1931. See also Appendix of this volume.
24 Recorded by Ghose, S., 'Tagore-Einstein conversation', 19 August 1930. See Appendix of this volume.
25 See note 21.
26 'My memories of Einstein', by Rabindranath Tagore. See Ghose (ed.), *The Poet and the Scientist*. Also see Chattopadhyay, D., *Rabindranath o Bijnan* (in Bengali) (Kolkata: Ananda Publishers Pvt. Ltd., 2000).
27 Ibid.
28 See note 21.
29 See note 23.

Appendix 1

The Tagore-Einstein conversation on the nature of reality

Note on the nature of reality

(A conversation between Rabindranath Tagore and Professor Albert Einstein on the afternoon of 14th July, 1930, at the professor's residence in Kaputh.)

T.: You have been busy hunting down with mathematics the two ancient entities, Time and Space, while I have been lecturing in this country on the eternal world of Man, the universe of Reality.

E.: Do you believe in the Divine as isolated from the world?

T.: Not isolated. The infinite personality of Man comprehends the Universe. There cannot be anything that cannot be subsumed by the human personality, and this proves that the Truth of the Universe is human Truth.

I have taken a scientific fact to explain this: Matter is composed of protons and electrons, with gaps between them, but matter may seem to be solid (without the links in the spaces which unify the individual electrons and protons). Similarly, humanity is composed of individuals, yet they have their interconnection of human relationship, which gives living unity to man's world. The entire universe is linked up with us (as individuals) in a similar manner: It is a human universe. I have pursued this thought through art, literature and the religious consciousness of man.

E.: There are two different conceptions about the nature of the universe: (1) The world as a unity dependent on humanity. (2) The world as a reality independent of the human factor.

T.: When our universe is in harmony with Man, the eternal, we know it as Truth, we feel it as beauty.

E.: This is the purely human conception of the universe.

T.: There can be no other conception. This world is a human world—the scientific view of it is also that of the scientific man. [Therefore, the world apart from us does not exist; it is a relative world, depending for its reality upon our consciousness.] There is some standard of reason and enjoyment which gives it Truth, the standard of the Eternal Man whose experiences are through our experiences.

E.: This is a realization of the human entity.

T.: Yes, one eternal entity. We have to realize it through our emotions and activities. We realized the Supreme Man who has no individual limitations through our limitations. Science is concerned with that which is not confined to individuals; it is the impersonal human world of Truths. Religion realizes these Truths and links them up with our deeper needs; our individual consciousness of Truth gains universal significance. Religion applies values to Truth, and we know this Truth as good through our own harmony with it.

E.: Truth, then, or Beauty is not independent of Man?

T.: No.

E.: If there would be no human beings any more, the Apollo of Belvedere would no longer be beautiful.

T.: No.[!]

E.: I agree with regard to this conception of Beauty, but not with regard to Truth.

T.: Why not? Truth is realized through man.

E.: I cannot prove that my conception is right, but that is my religion.

T.: Beauty is in the ideal of perfect harmony which is in the Universal Being; Truth the perfect comprehension of the Universal Mind. We individuals approach it through our own mistakes and blunders, through our accumulated experiences, through our illumined consciousness—how, otherwise, can we know Truth?

E.: I cannot prove scientifically that Truth must be conceived as a Truth that is valid independent of humanity; but I believe it firmly. I believe, for instance, that the Pythagorean theorem in geometry states something that is approximately true, independent of the existence of man. Anyway, if there is a *reality* independent of man, there is also a Truth relative to this reality; and in the same way the negation of the first engenders a negation of the existence of the latter.

T.: Truth, which is one with the Universal Being, must essentially be human, otherwise whatever we individuals realize as true can never be called truth—at least the Truth which is described as scientific and which only can be reached through the process of logic, in other words, by an organ of thoughts which is human. According to Indian Philosophy there is Brahman, the absolute Truth, which cannot be conceived by the isolation of the individual mind or described by words but can only be realized by completely merging the individual in its infinity. But such a Truth cannot belong to Science. The nature of Truth which we are discussing is an appearance—that is to say, what appears to be true to the human mind and therefore is human, and may be called *māyā* or illusion.

E.: So according to your conception, which may be the Indian conception, *it is not the illusion of the individual, but of humanity as a whole.*[1]

T.: The species also belongs to a unity, to humanity. Therefore the entire human mind realizes Truth; the Indian or the European mind meet in a common realization.

E.: The word species is used in German for all human beings, as a matter of fact, even the apes and the frogs would belong to it.

T.: In science we go through the discipline of eliminating the personal limitations of our individual minds and thus reach that comprehension of Truth which is in the mind of the Universal Man.

E.: The problem begins whether Truth is independent of our consciousness.

T.: What we call truth lies in the rational harmony between the subjective and objective aspects of reality, both of which belong to the super-personal man.

E.: Even in our everyday life we feel compelled to ascribe a reality independent of man to the objects we use. We do this to connect the experiences of our senses in a reasonable way. For instance, if nobody is in this house, yet that table remains where it is.

T.: Yes, it remains outside the individual mind, but not the universal mind. The table which I perceive is perceptible by *the same kind of consciousness which I possess.*[2]

E.: If nobody would be in the house the table would exist all the same—but this is already illegitimate from your point of view—because we cannot explain what it means that the table is there, independently of us.

Our natural point of view in regard to the existence of truth apart from humanity cannot be explained or proved, but it is a belief that nobody can lack, *no primitive beings, even.*[3] We attribute to Truth a super-human objectivity; it is indispensable for us, this reality that is independent of our existence and our experience and our mind—though we cannot say what it means.

T.: Science has proved that the table as a solid object is an appearance and therefore that which the human mind perceives as a table would not exist if that mind were naught. At the same time it must be admitted that the fact, that the ultimate physical reality is nothing but a multitude of separate revolving centres of electric force, also belongs to the human mind.

In the apprehension of Truth, there is an eternal conflict between the universal human mind and the same mind confined in the individual. The perpetual process of reconciliation is being carried on in our science, philosophy, in our ethics. In any case, if there be any Truth absolutely unrelated to humanity, then for us, it is absolutely non-existent.

It is not difficult to imagine a mind to which a sequence of things happens not in space but only in time, like the sequence of notes in music. For such a mind, such conception of reality is akin to the musical reality in which Pythagorean geometry can have no meaning. There is the reality of paper, infinitely different from the reality of literature. For the kind of mind possessed by the moth that eats that paper literature is absolutely non-existent, yet for Man's mind, literature has a greater value of Truth than the paper itself. In a similar manner, if there be some Truth which has no sensuous or rational relation to human mind, it will ever remain as nothing so long as we remain human beings.

E.: Then I am more religious than you are!

T.: My religion is in the reconciliation of the Super-personal Man, the Universal human spirit, in my own individual being. This has been the subject of my Hibbert Lectures, which I have called "The Religion of Man". In science we go through the discipline of eliminating the personal limitations of our individual minds and thus reach that comprehension of Truth which is in the mind of the Universal Man.

From "*The Religion of Man*" (Appendix II) supplemented from the typescripts in the Archive at Rabindra-Bhavana and *The Kenyon Review*, New Series, Vol. 23, No. 2, Cultures of Creativity: The Centennial Celebration of the Nobel Prizes (Spring, 2001), pp. 20–25, published by Kenyon College, which incoprporates the version that appeared in *The New York Times*.

Notes

1 In *The Religion of Man*, this phrase appears as 'it is not the illusion of the individual but of humanity as a whole'.
2 In the typescript at the Rabindra-Bhavana Archive and in *The New York Times*, the phrase appears as 'some kind of consciousness (which) we possess'.
3 In *The New York Times*, the phrase appears as 'not even primitive beings.'

Appendix 2

The Tagore-Einstein dialogue on youth, causality and music

(A conversation between Rabindranath Tagore and Professor Albert Einstein in the afternoon of 19th August, 1930, at the residence of a common friend, Dr. Mendel)

Einstein and Tagore

T.: What has impressed me this time in Germany is the work of the German Youth Movement. They are building up a community life on simple and large issues, not by reverting to medievalism but by applying the best gifts of the modern age to the service of a social ideal. Those splendid youths I saw in a Jugendherberge near Coblenz are not only managing their entire organization of lecture-tours, village work, social unions, but even building their own houses, cooking their food, bringing fuel from the woods and personally attending to the minutest detail of the daily life without depending on outside help for it. They are full of enthusiasm for the living of life itself, and every function of life therefore is joyous to them, rousing the feeling of adventure and self-expression.

E.: The Youth Movement we have in Germany today had, I believe, its origin in England.

T.: Yes, may be, but England has nothing like what you have in this country—you are so thoroughly earnest and vital in all that you do here that every new idea finds somehow its full velocity. In England the one institution which delighted me is Woodbroke near Birmingham, where the ideal of community life is maintained through social and educational activities related to our fundamental human aspirations. In Germany, however, as I have said, the vigour and the rich harmony of a complete life finds unique expression.

E.: Do these young men near Coblenz permanently live in that colony or is it a centre which they occasionally visit?

T.: Many of them live and work there and I believe, they also go out in batches on tours in their own country and outside. When once the centre is there, the element of adventure can also find its expression in works done elsewhere. It is this which makes their Youth Movement natural and progressive.

I was discussing with Dr Mendel today the new mathematical discoveries which tell us that in the realm of infinitesimal atoms chance has its play; the drama of existence is not absolutely predestined in character.

E.: The facts that make science tend toward this view do not say good-bye to causality.

T.: Maybe not; but it appears that the idea of causality is not in the elements, but that some other force builds up with them an organised universe.

E.: One tries to understand how the order is in the higher plane. The order is there where the big elements combine and guide existence, but in the minute elements this order is not perceptible.

T.: This duality is in the depths of existence—the two contradictions of free impulse and the directive will which works upon them and evolves an orderly scheme of things.

E.: Modern physics would not say they are contradictory. Clouds look as one from a distance, but if you see them near they show themselves as disorderly drops of water.

T.: I find its analogy in human psychology. Our passions and desires are unruly, but our character subdues these elements into a harmonious whole. Is it analogous to this in the physical world? Are the elements rebellious, dynamic with individual impulse and is there a principle in the physical world which dominates them and puts them into an orderly organisation?

E.: Even the elements are not without statistical order; elements of radium will always maintain their specific order now and ever onward just as they have done all along. There is then a statistical order in the elements.

T.: Otherwise, the drama of existence would be too desultory—it is the constant harmony of chance and determination which makes it eternally new and living.

E.: I believe that whatever we do or live for has its causality; it is good however that we cannot look through it.

T.: There is in human affairs an element of elasticity—some freedom within a small range which is for the expression of our personality. It is like the musical system in India which is not rigidly fixed as is the Western music. Our composers give a certain definite outline, a system of melody and rhythmic arrangement and within a certain limit the player can improvise upon it. He must be one with the law of that particular melody, and then he can give spontaneous expression to his musical feeling within the prescribed regulation. We praise the composer for his genius in creating a foundation along with a superstructure of melodies, but we expect from the player his own skill in the creation of variations of melodic flourish and ornamentation.

In creation, we follow the central law of existence, but if we do not cut ourselves adrift from it, we can have sufficient freedom within the limits of our personality for the fullest self-expression.

E.: That is only possible where there is a strong artistic tradition in music to guide the people's minds. In Europe music has come too far away from popular art and popular feeling and has become something like a secret art with conventions and traditions of its own.

T.: So you have to be absolutely obedient to this too complicated music. In India the measure of a singer's freedom is in his own creative personality—he can make permutations and combinations of notes according to the law of the melody prescribed but in all this he is guided by his own artistic conscience. He can sing the composer's song as his own if he has the power to creatively assert himself in his interpretation of the general law of the melody which he is given to interpret.

E.: It requires a very high standard of art fully to realise the great idea in the original music so that one can make variations upon it—in our country the variations are often prescribed.

T.: If in our conduct we can follow the law of goodness, we can have real liberty of self-expression. The principle of conduct is there but the character which makes it true and individual is our own creation. In our music there is a duality of freedom and prescribed order.

E.: Are the words also free? I mean to say, is the singer at liberty to add his own words to the melody which he is singing?

T.: Yes. In Bengal we have a kind of song—*Kirtan* we call it—which gives freedom to the singer to introduce parenthetical comments as he sings the song. This occasions great enthusiasm as the audience is constantly thrilled by some beautiful spontaneous sentiment freshly added to the song by the singer—phrases which had not been there in the original song itself.

E.: Is the metrical form quite severe?

T.: Yes, quite. You cannot exceed the limits of versification; in all his variations the singer must keep the rhythm and the time which is fixed. In European music you have a comparative liberty about time but not about melody, but in India we have freedom of melody with no freedom of time.

E.: Can the Indian music be sung without words? Can one understand a song without words?

T.: Yes, we have songs with unmeaning words—sounds which just help to act as carriers of the notes. We have two different types of songs in India. In Bengal poems and music combine with each other, in the north-west of India they have often songs with unmeaning or insignificant words. The idea is supplied by the music, the words remaining passive. In Bengal the words and the musical notes meet half-way and continue in the organic creation of a song.

E.: In Greece there has been like what you have in Bengal—music with poems.

T.: In North India music is an independent art—not the interpretation of words and thoughts. The music is very intricate and subtle and is a complete world of melody by itself.

E.: It is not polyphonous?

T.: Instruments are used, not for harmony but for keeping time [with] the music and for adding to the volume and depth.

E.: European music is not older than 400 years; it was only from then that music, polyphonous music, has been introduced.

T.: Has melody suffered in your music by the imposition of harmony?

E.: Sometimes it does suffer very much. Sometimes the harmony swallows up the melody altogether.

T.: Melody and harmony are like lines and colours in pictures. A simple linear picture may be completely beautiful—the introduction of color may make it vague and insignificant. Yet colour may by combination with lines create great pictures so long as it does not smother and destroy their value.

E.: It is a beautiful comparison; the line is also much older than the colour. It seems that the structure of your melody is much richer than here; Japanese music seems to be so.

T.: It is so difficult to analyse the effect of Eastern and Western music on our minds. I feel deeply moved by the Western music—I feel that it is great, that it is vast in its structure and grand in its composition. Our own music touches me more deeply by its fundamental lyrical appeal. European music is epic in character, it has a broad background and is Gothic in its structure.

E.: Yes, yes, this is very true.

T.: But this music is immense—I can never forget how much I was affected by its power once when a Hungarian lady played on her violin some pieces of music, both classical and modern.

E.: When did you first hear European music?

T.: At 17 I first came to know it intimately when I first came to Europe, but even before that time I had heard European music in our own household. I had heard the music of Chopin and others at an early age.

E.: There is a question which we Europeans cannot properly answer, we are so used to our own music. We want to know whether our own music is a conventional or a fundamental human feeling, whether to feel consonance and dissonance is natural or is it a convention which we accept.

T.: Somehow Piano confounds me. The violin pleases me much more.

E.: It would be interesting to study the effects of European music on an Indian who has never heard it.

T.: Once I asked an English musician to analyse for me some classical music and explain to me what are the elements that make for the beauty of a piece.

E.: The difficulty is that the really good music whether of the East or of the West, cannot be analysed.

T.: Yes, and what deeply affects the hearer is beyond himself.

E.: The same uncertainty will always be there, about everything fundamental in our experience, in our reaction to art, whether in Europe or in Asia. Even the red flower I see before me on your table may not be the same to you and me.

T.: And yet there is always going on the process of reconciliation between them, the individual taste conforming to the universal standard.

Berlin, 19 August, 1930.

Taken from the typescript in the archive at Rabindra Bhavana. (Reproduced in *The Kenyon Review*, New Series, Vol. 23, No. 2, Cultures of Creativity: The Centennial Celebration of the Nobel Prizes (Spring, 2001), pp. 25–30, published by Kenyon College, and also in *Asia,* March 1931, p 139.)

Contributors

Charles H. Bennett PhD, Harvard University, 1971; IBM Fellow, Thomas J Watson Research Center, York Town, USA; research interests: physical basis of information, quantum computing, cellular automata and reversible computing; with G. Brassard, discovered the concept of cryptography; one of the fathers of modern quantum information theory (cf Bennett's four laws of quantum information).

Harvey R. Brown Professor of Philosophy, University of Oxford (and an "unofficial" member of the Physics Department); Fellow of Wolfson College; studied physics at the University of Canterbury (New Zealand) and did his graduate studies in the Department of History and Philosophy of Science at Chelsea College (later merged with King's College) at the University of London. From 1978 to 1984, he taught in the Philosophy Department at the São Paulo State University at Campinas (UNICAMP), Brazil; co-winner of the 2006 Lakatos Award in Philosophy of Science and Fellow of the British Academy and President of the British Society for the Philosophy of Science.

Tian Yu Cao PhD in history and philosophy of science (1987), University of Cambridge, UK; Professor of Philosophy in the Department of Philosophy, Boston University, USA; Research interests: history and philosophy of Science, history and philosophy of contemporary physics (space-time, quantum field theory, particle physics, quantum gravity and cosmology).

Mihir Kr. Chakraborty PhD mathematics, former Professor of Pure Mathematics and Logic, University of Calcutta; research interests: non-standard logics, philosophy of mathematics, topology and functional analysis, writing on the interface of mathematics, philosophy, art and society. Currently a visiting professor at Jadavpur University; co-authored with Piero Pagliani a book titled *Geometry of Approximation*, Springer.

Avi Chaudhuri PhD from the University of California, Berkeley in 1987; postdoctoral studies at the Salk Institute for Biological Sciences and the University of British Columbia; Professor and James McGill Chair in Psychology, McGill University in 2002. Over the past 30 years, Chaudhuri has taught widely in both Canada and the United States in the areas of psychology, neuroscience, physiological optics and general biology. His textbook titled *Fundamentals of Sensory Perception*, Oxford University Press, 2011, has been highly acclaimed. Chaudhuri currently works as a consultant in the private sector.

Partha Ghose PhD from Calcutta University, 1968 under the supervision of Professor S.N. Bose; lecturer and then reader in physics, Visva-Bharati (1968–75); retired as professor and Academic Programme Coordinator at the S.N. Bose National Centre for Basic Sciences, Kolkata, in 1999; research interests: high-energy physics, foundations of quantum mechanics, classical polarisation optics, philosophy of science, science and music; published a book titled, *Testing quantum mechanics on new ground*, Cambridge: Cambridge University Press, 1999; at present, Honorary Scientist, National Academy of Sciences, India.

Ravi V. Gomatam Director, Institute of Semantic Information Sciences and Technology and the Bhaktivedanta Institute, Mumbai and Berkeley; adjunct professor at the Birla Institute of Technology and Science, Rajasthan; research interests: causal and ontological interpretation of quantum mechanics based on the application of the Schrödinger equation directly to the macroscopic phenomenal realm, independent of Born's probability rule and hence current statistical quantum mechanics. His work involves several ideas, including that of the ontology of 'objective semantic information' and 'relational properties' for matter. He has related research interests in macroscopic quantum computing, biology and semantic information as well as philosophy of language.

Dennis Lehmkuhl Research Assistant Professor at the California Institute of Technology and Scientific Editor of Albert Einstein's Collected Papers with the Einstein Papers Project at Caltech; studied physics and philosophy at the University of Hamburg and did graduate studies at Imperial College London, the London School of Economics and the University of Oxford; held positions at the University of Wuppertal (Germany), the University of Oxford (UK) and was a visiting fellow at the University of Pittsburgh (USA).

Kathleen M. O'Connell A scholar of comparative literature and modern Bengali culture; offers courses at the University of Toronto on Rabindranath Tagore and Satyajit Ray. A short bibliography: *Rabindranath Tagore: The Poet as Educator*. Second edition. Kolkata: Visva-Bharati. 2002, 2012. *Bravo Professor Shonku. Translation (Bengali to English) of three stories by Satyajit Ray*. New Delhi: Rupa & Co., 1985; 'Rabindranath Tagore: Facets of a Cultural Icon' issue, *University of Toronto Quarterly*. Edited by Kathleen and Joseph O'Connell. Vol. 77, Number 4, Fall 2008; *Rabindranth Tagore: Claiming a Cultural Icon*. Edited by Kathleen and Joseph O'Connell. Kolkata: Visva-Bharati, 2009.

Tushar K. Sarkar Premchand-Roychand Scholar; PhD from Calcutta University and University of Waterloo; research interests: analytic philosophy, philosophy of science, logic, Jainism; Director, Indian Council of Philosophical Research and Founder-Editor of the *Journal of ICPR* (1983–84); Founder-Coordinator of the School of Cognitive Science, Jadavpur University; Visiting Professor, Dept. of Philosophy, University of Waterloo (2000–2006); awarded *Acharya Tulsi Samman Puraskar* by the Jain Visva Bharati Institute (2001).

K. Sridhar Professor of Theoretical Physics, Tata Institute of Fundamental Research, Mumbai. Research interests: theoretical high-energy physics. He has recently completed work on a book entitled, *Particle Physics of Brane-Worlds and Extra Dimensions* for Cambridge University Press (in press). He has taught several courses on the philosophy of science and has co-edited a volume on Integrated Education entitled, 'Breaking the Silo: Integrated Science Education in India' (in press).

Anthony Sudbery A member of the Department of Mathematics at the University of York from 1971, retired in 2008; he is now a Professor Emeritus at York. His book *Quantum Mechanics and the Particles of Nature: an outline for mathematicians* was published in 1968, and he has published over 70 papers in mathematics, physics and philosophy.

C.S. Unnikrishnan Professor of Physics, Tata Institute of Fundamental Research, Mumbai; visiting professor at the Raman Research Institute, Bangalore; research interests: experimental and theoretical aspects of foundational issues in gravity and quantum physics, including quantum optics; key member of the IndIGO consortium and the LIGO-India project (www.gw-indigo.org) for setting up an advanced gravitational wave detector in India. He is a member of the LIGO Scientific Collaboration.

Index